I0473819

STATISTICAL MULTISCALING IN DYNAMIC ECOLOGY

Cover: Oxbow lake on the Tisza near Tiszafüred, Hungary

To Márta

I thank you for your tolerance, support, and lasting love

STATISTICAL MULTISCALING IN DYNAMIC ECOLOGY

Probing the long-term vegetation process for patterns of parameter oscillation

László Orlóci FRSC

London, Canada - 2012

 *Scada Publishing*

Refer to this book:

Orlóci, L. 2012. Statistical Multiscaling in Dynamic Ecology. Probing the Long-term Vegetation Process for Patterns of Parameter Oscillations. SCADA Publishing, Canada. Online edition: https://createspace.com/3830594

Look for:

Orlóci, L. 2013. Quantum analysis of primary succession. The energy structure of a vegetation chronosere in Hawai'i Volcanoes National Park. SCADA Publishing, Canada. Online Edition: https://createspace.com/4452597

Orlóci, L. 2013. Quantum Ecology. Energy structure and its analysis. SCADA Publishing, Canada. Online Edition: https://createspace.com/4406077

Orlóci, L. 2013. On the Energy Structure of Natural vegetation. In search for community governance rules. SCADA Publishing, Canada. Enlarged Online Edition: https://createspace.com/4153484

Orlóci, L. 2012. Self-organisation and Mediated Transience in Plant Communities. SCADA Publishing, Canada. Enlarged Online Edition: https://createspace.com/3585127

Orlóci, L. 2012. Statistical Ecology. The quantitative exploration of nature to reveal the unexpected. SCADA Publishing, Canada. Online Edition: https://createspace.com/3476529

Orlóci. L. 2011. Problem flexible computing in statistical ecology. SCADA Publishing, Canada. Online Edition: https://createspace.com/3574792

ISBN-13: 978-1475071382
ISBN-10: 1475071388
V 2014-01-01

Look for information:
https://sites.google.com/site/statisticalecology/

Statistical multiscaling
All rights reserved © 2012 László & Márta Orlóci
Laszlo.orloci@gmail.com

Expressions of gratitude regarding the inception and circumstances of the Essay: to Academicians Prof. Dr. Gábor Fekete (HAS and Zólyomi Bálint Foundation) and Prof. Dr. János Podani (Eötvös University), Drs. Katalin Török (Director, HAS Ecological and Botanical Research Institute, Vácrátót) and Marianna Bíró (HAS Ecological and Botanical Research Institute, Vácrátót) for hospitality and facilitation of the field trip to the Kiskúnság; Dr. Enikő K. Magyari (HAS) for discussions and her HAS data sets on loan for analysis; Mr. István Beliczay DFE for hosting the Sopron Memorial Forest visit near Tompa and presentation of a mock model of the 4th Century Hun settlement at Ördögárok; the Farkas family in Puszt-amérges for hosting the Kiss Ferenc memorial plaque dedication ceremony. The Essay benefited from attentive reading of an earlier version and suggestions by Prof. Dr. J. Podani and Prof. M. Mukkattu, and proof-reading of the definitive text by Márta Mihály DFE.

A message to students: *repetitio est mater studiorum*

Speed reading will not do. Technical terms are important; they should be understood and committed to mind early for fluent, thoughtful reading of the text. Have paper and pencil on hand, take frequent notes, and write critical reviews of the sections.

Contents

Contents.. 6
Preface 8
Perimeters of the case study 15
Further on the data... 18
The tangibles... 20
Fundamental conjectures 23
The analysis.. 25
   Broad outline... 25
   Basic concepts 27
      Trajectory trace 27
      Particle count transformations................................ 27
      Variable scale.. 28
      Variable time step width 28
      Parameter partitions 29
      Recursive analysis 29
   Elementary operations... 29
      Data blocking by pooling and averaging........................ 30
      Data smoothing by moving averages 30
      Equal time-step transformation 30

Tolerance radius 32
Homeomorphy 33
Shifting the record series to create a lag 36
Eigenmaps.. 37
Multiscale correlation 40
The *A* graph of Sarló-hát 40
Time delayed climatic effects... 42
Multiscale correlation analysis performed...................... 42
Interpretation of the ρ(A,T) graphs............................... 48
Interpretation of the Φ^- and Φ^+ graphs 51
What do these mean? 55
Indicator taxa 61
Process shape complexity... 73
Comparison of trajectories 76
Overview 81
Bibliographic references 85
Bibliographic supplements 90
Index 97
Biographical notes 100

Preface

I am presenting materials in this Essay from a short course which I taught to graduate students in the Ecology Program of the Universidade Federal do Rio Grande do Sul in Porto Alegre, Brazil. I should mention, familiarity with the notion of *multiscaling* was not a prerequisite for the course. The idea was developed from the ground up as if it were the first time the students heard about it. The course materials balanced theory and application. The Essay's contents emphasize the *modus operandi*.

Ecologists practice a form of multiscaling in their work when they use *successive approximation*[1] (Poore 1962, Wildi and Orlóci 1987, Orlóci 2001) to ascertain Nature's momentary states, yet the notion is not ecological, not even unique to Science. It is universal in the recognition that the observer's scale defines the observer's perception of Nature.

------

[1] In this, the next-step in data analysis very much depends on what has been discovered in previous steps. Refinements and reasoned changes in technique and direction are permitted at any step. Therefore, the conclusion condenses around itself within increasingly smaller radius. When it cannot be further condensed with the expenditure of reasonable effort, the conclusion becomes an inference.

To gain a first approximation to the definition of multiscaling as an analytical process, it is sufficient to perform a simple experiment. Two things are needed for this: a digital camera and a well-lit object. At this time the object is the Oxbow landscape in the Tisza floodplain near Tiszafüred. Two photographs were taken from the same position with different magnification:

As the power of resolution increased, so did the identifiable detail in the picture. The important thing to remember is the fundamental generalisation: the object remains the same, only the viewer's perception of the object changes owing to the *scale effect*.

The experiment reveals a simple procedure in *multiscaling*. The same method is repeated at different scales, in this case with different settings of the power of resolution in the camera. There could be extensions to the experiment if the telemetric data permitted us to go that way. Post processing[2] of the electronic files by some pixel metric application program could be done to sharpen the image, change the colours or bring out further high-resolution details.

The phrase "if the telemetric data permitted" is important. This implies that the data base may or may not hold the extra information needed to improve the image for better perception of the component objects. In

---

[2] In the present case, the One Step Photo Fix function in Corel's PaintShop application program improved sharpness in each photo images. The One Step Noise Removal function destroyed the image in each case by what appears to be the removal of Nature's fine finishing brushstrokes that give the pictures quality.

other words there may be nothing of interest left to look for in the tele-metric data.

What is then *statistical multiscaling*? We can answer this question on a very general level if we expand the purpose of the experiment into the realm of pattern recognition. The latter invites this question:

Could the spatial arrangement of objects we identify in the photograph be the outcome of a purely random spatiotemporal process?

The question brings up an interesting answer. It is absolutely certain that the way Nature painted the Oxbow landscape by random strokes is very different from the pictures the camera has taken. But which of the elements from Nature's original stayed and which have been lost? It is quite correct to assume that the plants we see are the survivals of those that came at random to the site, but the ones we see are only those remaining after self-organization in the community and environment mediated transience has done their work (Orlóci 2011). So in this more complex context the question invites a statistical answer and by its virtue the answer has to be probability based.

It should be evident to the reader that multiscaling implies the making of the scale a variable in statistical analysis. The scale itself does not become a target of the analysis; it is a facilitator of effective analytical probing of data sets for intrinsic details not apparent on cursory inspection of the phenomenon we are observing.

A common aspect of statistical multiscaling is the exact repetition of the same analytical steps (algorithm) after each change of the scale. This has its consequences. One of the more obvious ones is the outcome, which is no longer a single estimate of the parameter value in question, but a set of as many independent estimates as there were scale changes in the multiscaling process.

The idea of multiscaling in the above sense is not new in ecology. My post-doctoral supervisor Peter Greig-Smith has used this technique more than six decades ago in the analysis of coincidental ground patterns of plant populations and environmental conditions (Greig-Smith 1952). The Greig-Smith technique allows the user to pinpoint the characteristic geodesic scale at which the pattern of a specific population coincides most clearly with the pattern of an environmental condition. The 1952 technique flourished and proliferated into many variants. I refer the reader for further readings on this to Greig-Smith's own summary which he makes in the 3rd edition of *Quantitative Plant Ecology* (1983).

The original Greig-Smith technique uses sampling unit size for scale. Changes in the scale involve passing from the basic unit size $u=1$ to larger unit sizes by concatenation, usually but not necessarily in the manner of $u = 1, 2, 4, 8, ...$. The performance parameter for all variables is the *partial variance* for which a new value is computed for each value of u. Ecological conclusions are drawn about species pattern response and environmental forcing from the comparative analysis of peak displacements or peak coincidences in the partial variance graphs. Partial sums of squares, logarithmic expressions of proportions, and other types of performance parameters are used by others with the same or similar objectives. Examples are found in an early paper by Orlóci (1971) and many more in Greig-Smith's book (1983).

The Greig-Smith technique has served as a model in the development of new methods in Quantitative Ecology. Examples include the analysis of phytosociological (contingency) tables (Feoli & Orlóci 1985), sample size optimization in ecological surveys (Orlóci & Pillar 1989), vegetation edge detection on transects (Orlóci & Orlóci 1990), and multiscale trajectory analysis (see Orlóci 2009, 2010) - just to mention a few. The references in these should lead the reader to others.

In our specific case, the *objects of statistical multiscaling* are palynological spectra. The scale variables include time step width, sliding window size, tolerance radius length, confidence belt width, and others as appropriate in specific instances. The performance parameters are different types and always pertain to a characteristic of the compositional transition process. The Essay imports performance parameters from three branches of mathematics: Newtonian motion dynamics, Mandelbrot fractal geometry, and Eulerian topology. Included are such well known functions as compositional transition acceleration (deceleration), process shape complexity, and process homeomorphy (homotopy). An important point not to be missed from here on: the fundamental parameter oscillations we are considering are subordinated to time.

Using the familiar terms of biostatistics, multiscaling takes us from estimation of a Nature state to estimation of scale dependent Nature state vectors, as many in number as there are scale variables. It is to be emphasized that the object of the analysis remains the oscillation of the same parameters of the same process, only the visibility of the structural details keeps changing under the fundamental scale changes. I re-emphasize: the same statistical analysis of the same data set is performed at the basic scale and repeated after each change of scale. This allows intrinsic[3] details to come into view and to be seen evolving in space-time.

We can only wonder what direction the science of Ecology would have taken in the early 20th Century if the starch antagonists, H.A. Gleason and F.E. Clements in particular, considered the vegetation process a scalable object and saw its perception as a consequence of scale. They had precedent for this more highly evolved view of Nature in the anecdotal accounts of Anton Kerner von Marilaun's monograph as far back as in 1863.[4]

---

[3] "intrinsic" in Collins Dictionary: ... belonging to something as one of the basic and essential features that make it what it is ... . 'Hidden' is not a valid synonym, but often misused as such. essential features that make it what it is ... .

[4] See details about the antagonists in the wonderfully complete, exquisitely written monograph of Robert P. McIntosh (1985) on the subject. Regarding Kerner (1863), the reader may fall back on Conard's (1951) English translation.

Multiscaling in dynamic ecology

The examples in the present Essay revisit a well-worked topic in paleoe-cology, the Holocene vegetation history of NE Plain in Hungary. Fasci-nated by the available results, by insight gained from published sources and on field trips, and by the potential of Enikő Magyari's HAS data set[5] led to my decision to offer the use of a *fresh* methodology for the *next step* in successive approximations of the Holocene vegetation history of Hungary's NE Plain.

The methodology emphasises the compositional transition process in palynomorph spectra. Conceptually and methodologically the multiscal-ing approach is steps removed from the techniques concerned with the tangibles, such as the entering and dropping out of taxa, and the use of plant indicators selected on the basis of current geographic distributions to identify past climate types.

The objective of our process oriented analysis is the detection of mul-tiscale response patterns in the parameters and to identify the levels of their dependence on specific types of environmental forcing. What did I find out in the way of generalizations, after having performed immense amounts of code writing and ensuing computations about the process properties? In a nutshell:

i. The compositional transition process generates a complex, self-similar, hierarchically embedded pattern in time.

ii. Linear process phases are punctuated by hotspots of violent composi-tional perturbations in delayed synchrony with oscillations in the global atmospheric temperature.

iii. More humid and more arid climates are separated by hotspots of compositional transitions.

---

[5] Dr. Enikő K. Magyari (HAS) lent me her HAS data sets for analysis from two sites in NE Hungary and from one site in the Retyezát Mountain.

13

iv. The shape of the process trajectory measured in fractal geometric terms is complex, yet process determinism is very strong, notwithstanding its convolution with a sizable random component.

v. The von Post doctrine of regional process parallelism is upheld.

The technical message of the Essay is substantial. It fixes the path of statistical multiscaling from theory to worked examples. But there are requirements for reliable results. These include long palynological spectra (or other type of ecological sere) and high intensity sampling of the sediment core (or transect), particularly when statistical estimation and significance tests are intended.

As it started out, the Essay's main objective should be the presentation of the multiscaling method in a matter of fact way through examples. This seemed practical enough and also sufficient under the circumstances to show the powers of statistical multiscaling in a concrete case, for which results were available from other palynological studies. But it did not work. It became clear as the writing progressed that the interpretation of the results is far more complex than anticipated. Intended users needed much additional details about the theory and more explanations of the technique within the Essay itself. Time will tell if the expanded contents are sufficient. But then the user can visit the references for further details.

L. Orlóci
Kailua, Hawaii -- March 15, 2012

Perimeters of the case study

The context in which we apply multiscaling to the Magyari data sets is definitely Statistical Paleoecology in practice. The data sets include real palynological spectra from three sites: Sarló-hát (SH) and Báb-tava (BT) on the NE Hungarian Great Plain, Taul Dientre Brazi (TDB) in the Retyezát Mountains of the Carpathian Range. The spectra are high precision for sampling and taxon identification. The data sets contain dated and calibrated particle counts of 122 taxa from 94 sediment strata in SH, 114 taxa from 75 strata in BT, and 134 taxa of 91 strata in TDB. The period length extends from 11380 to 80 BP[6] in the SH spectrum, 7780 to 1138 BP in the BT spectrum, and from 15755 to 9956 BP in the TDB spectrum. We focus on the SH spectrum and use the BT and TDB spectra for reference in statistical comparisons. Note that the strata dimensions will change after equal time step transformation to 226, 134, and 116. We

---

[6] Important note: in most cases the dates we given were given to us, read by us from graphs, or found analytically. We take the reading as they come but do not claim more precision than customary with dates in palynology in general. Measurement errors in the isotope dates are not available to us.

will return to the topic in the sequel.

Magyari et al. (2008) characterise the regional climate of the NE Plain as 'transitional', warm to cold, continental. Szász & Tőkei (1997) is their source for meteorological data:

-- mean annual temperature *9.4–9.5⁰C*,
-- mean January temperature *-3.5⁰C*,
-- mean summer temperature *20.5⁰C*,
-- annual precipitation range from *630* to *660* mm,
-- predominant wind direction northerly, south-easterly and south-westerly winds less frequent.

I should mention another relevant fact about the climate. According to Borhidi (1961) the Great Plains's climate has the characteristics of the semiarid Pontic steppe climate of the Black Sea's west coastal region in 20 out of 100 years. I suppose more often the climate is more like on a forest step, perhaps an Oak Savannah (Fekete et al. 2008, 2010).

I use two environmental data sets in the example. The historic atmospheric temperature records are taken for the relevant periods from the Vostok temperature differences series as presented by Petite et al. (2001)[7]. Orlóci (2008) discuss temperature transformations and argues in favour of the suitability of the Vostok records for use in long-term ecological studies outside the Antarctic region.

---

[7]As far as the estimation of historic atmospheric humidity is concerned, the classical approach starts with the geographic distribution of species on current climatic gradients to determine species indicator values. The indicator value of a species is taken to be inversely proportional with the breadth of its distribution on the climatic gradient. In the next step the vertical distribution of the indicator species is mapped onto the palynological spectrum and on that basis the periodic climatic conditions are postdicted. A most éclat example of this approach, with direct relevance to the Magyari sites, is the one we find in Zólyomi Bálint's (1936) work on the Holocene vegetation history of Hungary. I emphasize that the method relies on current distributional limits of species as proxy for the temporal distribution limits of palynomorph taxa.

The 2001 version of the Vostok temperature series has 341 measurements within the *0* to *15770* yr BP period. The bottom of this segment of the series is at 340 m deep in the Antarctic Ice Shield. This implies on average about 1 temperature measurement per 46 years.[8]

Another of the environmental data sets is from estimates of atmospheric humidity at historic time point in terms of the relationship of the Thornthwaite index (*Th,* see Trewartha 2001) to atmospheric humidity on a broad regional basis. The proxy equation of the relationship is discussed by Orlóci's (2008). We return to this topic later in the text.

Available results from Magyari et al. (2001, 2001b, 2002, 2008b) tell a compelling story of the climatic effect on the vegetation history of the region. It is quite clear from our results as well that the current climate follows on the heels of a definitely more arid climate in a historic chain of more climatic aridity to more humidity through the Holocene. We know from our results in this Essay that the latest transition, beginning around *1700 BP,* came with much turbulent change in the palynomorph composition of the SH plant particle spectrum, without any lasting subsidence by as late as *80* years before the core samples were taken. But this last major turbulence should not be attributed to the changing climate alone. The effect of massive grazing, that gained momentum with the arrival of the Huns in the late 3rd Century AD, and the Magyars (Hungarians) in the 8th Century, large scale land cultivations and flood controls on the Tisza since the 18th Century, should be substantial.

---

[8] We use Vostok temperature differences in this Essay. The differences are measured from a common base provided by the temperature equivalent of the current deuterium content in the oceans. Deuterium is a stable isotope of H. The Vostok series spans more than four complete glacial periods. Refer to Petite et al. (2001) for additional technical details. The global utility of the Vostok temperature records in palynological studies has been shown by Orlóci (2008, 2009 and references therein) mainly based on the general results of Swine Gruber (1996).

Further on the data

It should be expected that paleoecological data will carry a heavy load of random variation. It is important to see clearly that random variation is intertwined with determinism (directed variation) which we seek to unveil by multiscale statistical analysis. Where does random variation come from? It is the consequence of chance events inherent in the functioning of plant populations and the environment, in site selection, in the method of sediment sampling, in plant particle type identification, in the technicalities of sediment dating, and quite possibly in many other things related to and having influence upon the data and its analysis.[9] What random variation does not include are the barefaced mistakes committed by persons.

What ever is the source for random effects, it is imperative that statisti-

---

[9] For detailed discussions of error sources and anticipated effects see Orlóci et al. (2002, 2006), Orlóci (2009, 2010 and references therein).

cal data analysis involves extra steps which isolate the total random effect in order to make determinism seen at its best.

The techniques of multiscaling help us to isolate and measure the magnitude of the total random effect. Related to this, we define a null state to which we can compare any observed state. In the definition of the null state we assume the complete rule of chance over compositional transitions in the palynological spectrum. Such a definition of the null state has context and utility in Monte Carlo experiments which are our principal tools to construct probability distributions and to determine statistical expectations in process parameter oscillations. [10]

---

[10] Reference is made to Hammersley and Handscomb (1964) for the theory of Monte Carlo techniques and to Orlóci and He (2009) for example and references to the application of Monte Carlo experiments in ecology and the related fields under the less preferable epitaph of randomization testing.

The tangibles

Palynological spectra (like in Fig. 1) and atmospheric variables are the tangibles on which we perform statistical multiscaling. What makes a variable tangible? -- the records that describe its variation.

The building blocks of a palynological spectrum are plant particle counts in sediment samples taken from selected depths. We refer to the plant particle types as *palynomorph taxa*.

Regarding the atmospheric variables, we have access to historic temperature and humidity data (see preceding section). We consider the anthropogenic effect determining in its importance, but without hard data its effect remains to us no more than anecdotal history.

The focus of the analysis remains the compositional transition process, specifically targeting the parameter oscillations encapsulated within palynological spectra. But these are now imbedded within a broad discourse on foundations and technical details.

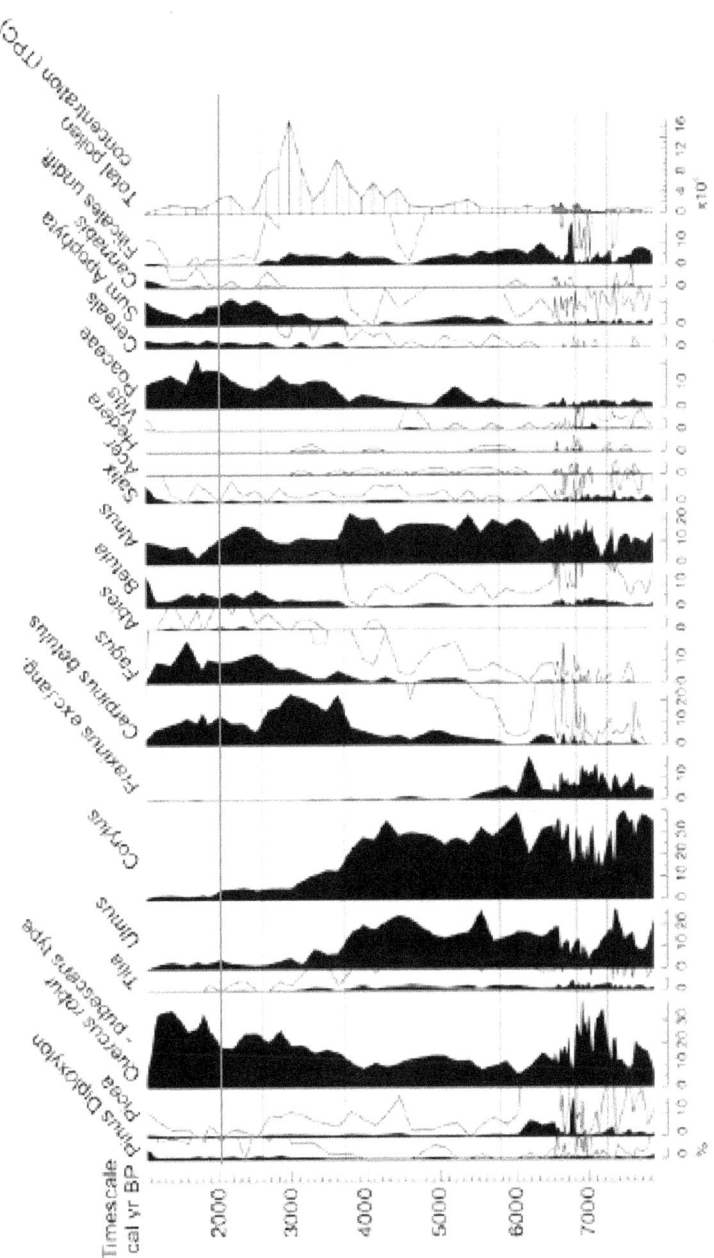

Fig. 1. Partial palynological spectrum adapted from Magyari et al. (2008, Fig. 4, p. 39). Origin: Báb-tava, NE Hungary. Relative frequencies are shown (% scale) for selected palynomorph taxa. Each point (*e.g.* 2000 yr

BP) on the time scale has an associated state vector marked out by the line drawn through the time point, such as in the case of year 2000 BP. The elements of the state vectors are the palynomorph frequencies intercepted in the individual taxon graphs. We refer to such a state vector as a paleorelevé. The light lines shown delimit the Magyari et al. (2008b) local pollen assembly zones. See further details regarding the data in the original publication.

.

Fundamental conjectures

Three fundamental conjectures define the exact analytical tasks for the Essay:

CONJECTURE 1. Historic hotspots of compositional transitions in the spectra are in delayed synchrony with atmospheric temperature and humidity oscillations. Can this be tested? We suggest an indirect way by proxy.

CONJECTURE 2. The long term vegetation process has well defined determinism braided into much randomness. Can the relative importance of the two components, the directed and the random, be isolated? Our approach defines the random component as a blurring effect on determinism and uses techniques to suppress the blurring in steps of scale changes.

CONJECTURE 3. The von Post doctrine (1946) of regional process parallelism, a climate forced process homeomorphy (process co-ordination) does exist. The lack of uniform time step width and common natural starting point of the process in different sites complicate the testing. We suggest resolution of the complications by time step transformations (see technique in next section).

The analysis

Broad outline

The generic name for the methodology we are using is *multiscale trajectory analysis* (Orlóci 2009, 2010,2010b) [11,12]. By this methodology we put the compositional transitions in palynological spectra onto the analytical

---

[11]Orlóci's webpage www.vegetationdynamics.com defines trajectory analysis in the context of long-tem vegetation dynamics. The definition emphasizes that vegetation dynamics is scalable based on the characteristics of the *process trajectory*. The trajectory is the imaginary line traced out in time by vegetation transitions. The definition observes further that the trajectory line is symptomatic of factor influences and governing principles, and because of this it should be an object of central interest in dynamic ecological studies.

[12]For information on the availability of the publications, the reader should visit https://sites.google.com/site/statisticalecology or http://www.vegetationdynamics.com/

pallet. The analysis hands us estimates of motion, fractal, and homeo-morphic parameter vectors. The vectors are presented in the form of graphs such as the compositional transition acceleration (deceleration) or *A* graph, process fractal dimension or *D* graph, and homeomorphic coefficient or *TC* graph. The *A* graph is specifically suited to locate historical hotspots of change; the *D* graph helps with highlighting the magnitude of the trajectory's shape complexity, or equivalent, the strength of the total random effect; and the *TC* graph puts the interpretation of the scale dependence of homeomorphy into comparable terms. Using the graphs, we can characterise the transition process in terms of phase structure, complexity, levels of self-similarity, co-ordination of duplet palynomorph spectra, and so forth.

The first step in the analysis is the construction of a parsimonious map of the process trajectory in phase space. The axes of the phase space are the palynomorph taxa. The time axis is the trajectory line.

Our mapping technique is the algebraic manoeuvre known as Eigenanalysis[13]. We discuss the reasons later in the text why this mapping technique is the only practical alternative available for us, unless the exact functional form of the trajectory is given *a priori* (e.g. a Markov chain).

What do we do with the trajectory map? We perform analyses on it with end-results including the numerical values of the trajectory parameters and other emergent quantities. On the basis of these values we can diagnose the intrinsic properties of the compositional transition process, the extrinsic forcing factors, their temporal and spatial patterns, hotspots of change, and the strength of homeomorphy.

---

[13]Eigenanalysis is the mathematical core of Principal Components Analysis, a broadly used ordination technique in Ecology. "Eigen" translates into English as efficient, characteristic.

Basic concepts

Trajectory trace

The trace is an imaginary line through a set of points within phase space. Each point is associated with a data vector to which we referred as a *paleorelevé* in Fig. 1. The paleorelevé fixes the compositional state of the palynomorph assembly (palynomorph community, palynomorph collection) at a given time point. Its elements are palynomorph particle counts.

Particle count transformations

We may take the particle counts as given to us by a competent pale ecologist and perform Eigenanalysis on the co-variance or cross products matrix (Orlóci 2010). Either of these matrices will preserve the same structure and structural connections in the data undistorted, except for a factor *(n-1 or n)* when the covariance matrix is used. We may choose to transform the particle counts before Eigenanalysis. But users must be aware: if chosen in ignorance of the potential consequences, a data transformation can distort the original data structure and make multiscale trajectory analysis pointless.

We do a simple transformation from rectangular pixel co-ordinates to polar co-ordinates on the Oxbow photograph which we already used in the Preface to illustrate the point:

The transformation which produces the transformed image on the right has two steps. In the first we convert the x and y rectangular co-ordinates of any pixels to polar co-ordinates ρ and ϑ relative to the polar axis,

$$\rho = \sqrt{y^2 + x^2} \text{ and } \theta = \arctan(y/x)$$

We add π to ϑ when $x < 0$ and $y > 0$ or we subtract π from ϑ when $x < 0$ and *y<0.* To create the image on the right hand side, we use ρ and ϑ as rectangular co-ordinates. Clearly, a simple appearing transformation, such as as ρ for *y* and ϑ for *x,* can result in a substantial distortion of the data structure. In the Oxbow example the transformation rearranged the pixels into circular bands from right top anticlockwise back to the right top. It is, however, important to see that the distortion is not haphazard. This makes the distorted image a unique signature of intrinsic properties in the original image.

The transformation done, but the question remains: which pair of the performance scalars should we subject to statistical multiscaling. Should it be *x,y* or ρ,ϑ. The user's choice is relatively simple when images are seen placed side by side. In most cases this is not so. The consequences of the transformations have to be reasoned out from basic principles or determined by experimentation.

Variable scale

We already mentioned that a unique aspect of multiscaling, which separates it from uniscale statistical analysis, is the making of the scale a variable. This applies to scales of any kind. It is by this approach that conventional statistics has its link to Poorean successive approximation (Wildi and Orlóci 1987, Orlóci and Pillar 1989, Pillar and Orlóci 1993, Orlóci 1993, 2001, 2001b, 2010).

Variable time step width

Time step width, the elapsed time between neighbour trajectory points, is usually irregular. This is a necessary consequence of sampling which supplies material samples for dating. Data transformation to equal time step width is needed to facilitate comparisons or to change the resolving power of the observations. As we have seen already, time step transformation can lead to data smoothing and reduction of the random effects. All parameters are likely to be affected.

Parameter partitions

Clever analytical partitions allow us to perceive hierarchical parameter structures in the ecosystem. The partitions for best results have to be additive. There are two generically different ways to do additive partitions, logical or analytical. Two examples --

Example 1: $m = m_E + m_I$ This is a logical partition when we define m as the performance of a plant, m_E as the current environment mediated effect, and m_I as the consequence if inheritance. The analytical handling of such a partition is explained in Orlóci (2011).

Example 2: $V = V_1 + V_2 + V_3 + ... + V_s$ This is a purely analytical partition valid if we think, for example, in terms of s orthogonal functions each specific to a given species taken in order from $i=1$ to $i=s$ as they appear in a relevé. Letter V stands for the total variance in the sample of s species and V_i is the variance of the *ith* orthogonal function. V_i is called the residual variance of species I in the list. The residual variance of a species depend on its position in the list and V_1 is the true variance of the first species. Such an enumerative dependence of the residual variance is a peculiarity of partitions based on orthogonal functions (see Orlóci 2010).

Recursive analysis

To take advantage of the scale's synonymy with the analysis' power of resolution, a complete statistical analysis is performed at each resetting of scale values. But then, the result is, as we already mentioned, not a set of single parameter states, but a set of parameter state vectors.

Elementary operations

We discuss the elementary operations the easy way of simple examples where it is possible. In some cases we need to fall back upon theory or at least make brief reference to sources where the theory is presented.

Data blocking by pooling and averaging

Each observation within a block is replaced by the block average. Data smoothing occurs and intensifies with block size.

8-valued time series at time step *1TS*: 5 3 7 6 1 2 4 8
8-valued series at step size *2TS*: 4 4 6.5 6.5 1.5 1.5 6 6
8-valued series at step size *3TS*: 5 5 5 3 3 3 6 6
etc.

Data smoothing by moving averages

What moves is a window, the "sliding windows" covering a block of data elements.

8-valued series at block size *1BS*: 5 3 7 6 1 2 4 8
8-valued series at block size *2BS*: 4 5 6.5 3.5 1.5 3 6 8
8-valued series at block size *3BS*: 5 5.33 4.67 3 2.33 4.67 6 8
etc.

Equal time-step transformation

We selected *50 ^{14}C yr* for equal time step width, symbolically *NSW*, in the examples. to separate the observed relevé points on the trajectory. This happens to be smaller than any inter-point distance in the trajectory map.

Our equal time step transformation algorithm performs seriation first by laying *1 ^{14}C yr* time units contiguously on the trajectory line and then summing the units within each *50 ^{14}C yr* period. The basic recursive step equation in the algorithm for any new time step width has the form $X_{jt} = X_{j0} + t(d_{jk} / T_{jk})$ where $d_{jk} = X_j - X_k$ and *t = 0, 1, ..., T_{jk}*. In this *j* and *k* are neighbour points (paleorelevés) on the trajectory, T_{jk} is the observed time step width between trajectory points *j* and *k*, and $X_{j0} = X_j$. This means that each relevé point is the beginning of a new segment in the seriated trajectory of *L* points. *L* is the total period length of the trajectory. The total length of the seriated trajectory containing n

paleorelevés is $T = \sum_{j=1}^{n} T_j$.

After seriation is complete, the elements are summed in blocks of the uniform time step width units to give us the equal time step trajectory of T/NSW relevé points.

The following example uses a simple data set suitable for longhand calculations to learn the technique:

```
Sample time series TBP: 1   8    12   26   30   yr
Sample time series data X: 2   4   8   10   15
Number of variables (palynomorph taxa) in X: 1
Number of observations per variable: 5
Character total: 39

Stretched   series   (hint   for   second   element,   X+t(d/T)=
2+1(2/7)=2.2857143):
2    2.2857143    2.5714286    2.8571429    3.1428571    3.4285714
3.7142857    4   5   6 7   8   8.1428571    8.2857143    8.4285714
8.5714286    8.7142857    8.8571429    9 9.1428571    9.2857143
9.4285714   9.5714286   9.7142857   9.8571429   10   11.25   12.5
13.75   15

Series total: 227.5
Number of cells in stretched series: 30
Step size chosen for condensation: 3
Unadjusted data out (hint for first element, 2 + 2.2857143
+ 2.5714286 =6.8571429):   6.8571429   9.4285714    12.714286
21    24.857143    26.142857    27.428571 28.714286    31.107143
41.25
Adjusted data (hint for first element, 6.8571429/(39/227.5)
= 1.1755102):
1.1755102   1.6163265   2.1795918   3.6   4.2612245   4.4816327
4.7020408
4.922449   5.3326531   7.0714286
Adjusted time points TBP: 1   4   7   10   13   16   19   22   25
28
Output vectors: 1 taxon, 10 valued
```

An application program to calculate equal time steps automatically is included in the external appendix of Orlóci's (2010) Statistical Ecology.[14]

Transformation of the time steps to equal width is one of the remedies we use to establish the comparability of palynological spectra from different sediment cores. The effect of equal time step transformation is data smoothing -- and definite information loss after the *NTW* exceeds the smallest time step in the original data. What should be considered important is the scaling up to lower powers of resolution which is expected to cut into chance compositional variation (smoothing effect), and by loss of such information a clearer view of the core determinism of the compositional transition process may be gained.

Tolerance radius

Fig. 2 has the basic graph on which we can explain what a tolerance radius is and how the tolerance radius controls decisions regarding the analytical distinguishability of the trajectory points. The graph is a map of the first four points (**A, B, C, D**) of a trajectory in palynomorph phase space. Two palynomorph taxa are the axes (X_1, X_2). Point **A** marks the observed state farthest in the past. Each subsequent point (**B, C, D**) is progressively closer to present.

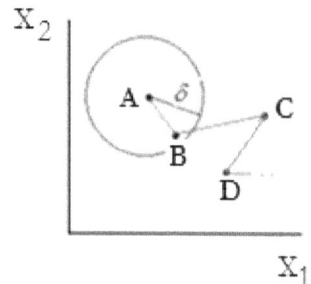

Fig. 2. Phasespace map of the first four points of a trajectory. The phasespace is a reference space whose axes are the palynomorph taxa. The palynomorph taxa are plant particle types which make up a palynological spectrum. The trajectory is the time axis of phasespace. Letter δ is the tolerance radius.

We assume that equal time steps separate the trajectory points. So the time elapsed between states **A** and **B** is the

[14] For availability visit https://sites.google.com/site/statisticalecology/

same as the time separation of **B** and **C**, and **C** and **D**. But the lengths of the line segments $\overline{AB}$, $\overline{BC}$ and $\overline{CD}$ are different. The length of the line segments is proportional to the velocity of compositional transitions when computed in full-dimensional phase space.

The *tolerance radius* δ is a scale variable. It takes on values 0, 1, 2, 3, ..in time step units. What is accomplished by the tolerance radius? Its magnitude decides whether given process states fixed by paleorelevés are or are not deemed analytically distinguishable. For example, at the given δ, process state **B** is deemed indistinguishable from process state **A** but both are distinguishable from process states **C** and **D**. When δ equals the distance of the two most distinct states, the most distant trajectory points, no state will be deemed distinguishable. Clearly, the manipulations which increase the value of δ are analogous to data smoothing and in that sense to reduction of the random effect.

Homeomorphy

We return once again to the four points **A, B, C** and **D** in Fig. 2. The phasespace origin is the analytical zero point corresponding to $X_1 = X_2 = 0$. In the case study to be discussed the number of trajectory points (process states, paleorelevés) and also the number of X variables (palynomorph taxa) can run into the hundreds. To explain our scoring system for the purpose of determining the homeomorphy of the trajectory in Fig. 2 with another as yet unspecified trajectory, we note the following:

i. Whereas the quadruplet **A, B, C, D** represents a time series from past toward present as a straight sequence from **A** to **B** to **C** to **D**, the trajectory mapping itself is not a simple strait line, but an irregular polygon. In the example at each point the trajectory changes direction.

ii. The first reference point in the scoring system is **A**. This point maps the paleorelevé of the deepest examined horizon of the sediment core, into phase space. The corresponding TBP age is where the analytical time begins.

iii. The scoring system's first reference direction and reference distance is marked by the first trajectory segment $\overline{AB}$.

iv. Points **A** and **B** are deemed identical at the given tolerance radius δ since **B** is within the tolerance sphere of **A**. Therefore, we score zero (0) for the **A** to **B** comparison. Should **B** be a point outside the tolerance circle then we would score a plus (+).

v. Observing that the inner angle at point **B** is greater than 90° and the length of the $\overline{BC}$ segment is larger than δ, we score for the **B** to **C** comparison a plus (+). Should the angle at **C** be acute (<90°) then the appropriate score would be a minus (-).

vi. We choose trajectory segment $\overline{BC}$ for the new reference direction and continue scoring the **C** to **D** comparison. The inner angle at **C** is acute and the length of $\overline{CD}$ is greater than δ. For this reason the **C** to **D** comparison receives a minus (-) score.

With more than four points, we would choose the $\overline{CD}$ segment for our new reference direction and continue scoring in the same manner until the last (most recent) trajectory point is reached.

Homeomorphy is measured between two trajectories. We label the case in Fig. 2 as Trajectory *I*. Exactly the same scoring technique is performed on the second trajectory, labelled Trajectory *II*, with uniform time step width the same as in Trajectory *I*. The time interval covered in the two trajectories is the same.

We have then two score vectors $SV_I\delta$ and $SV_{II}\delta$ for tolerance radius δ:

$$SV_I\delta \quad SV_{II}\delta$$

0	+
+	+
-	-

Now, we compare element by element the two score vectors. When the corresponding scores are identical such as 00 (none in the table), -- or ++, an elementary homeomorphy is detected and the counting variable *M* is incremented by 1. After all comparisons completed at the given tolerance radius δ, we calculate the topological coefficient:

$$TC_\delta = \frac{M_\delta}{n} = \frac{2}{3} = 0.67$$

TC_δ is in fact a measure of the relative intensity of homeomorphy between the two trajectories at tolerance radius δ, *n* is the number of score pairs.[15] In real cases we assume a large *n*. TC_δ is measured on a zero to one scale with random expectation $e_\delta = 0.5$ at $\delta = 0$. Note that e_δ changes with δ.

When TC_δ is not significantly different from e_δ then we conclude that at least one of the trajectories is generated by a random compositional transition process. Since the break point is e_δ and not 0, values of TC_δ significantly greater than e_δ indicate homeomorphy more intense than the expected. When the TC_δ value is significantly less than e_δ then homeomorphy is considered less intense than the expected.

We can determine probabilistic (confidence) limits around TC_δ in an ap-

---

[15] The *00* scores should not be left unrecorded. They are required for comparability of the $TC_\square$ for different values of $\square$.

propriately designed Monte Carlo experiment. The Monte Carlo experiment's basic assumption is that the expectation of TC_δ is e_δ. In other words, chance is reigning over compositional transitions in at least one of the spectra. The lower and upper limits are used for testing the significance of the observed TC_δ values.

A complete example is given later. The scale variable δ is taken through its range and the resulting TC_δ values, their expectations e_δ and statistical upper *(UL)* and lower *(LL)* confidence limits are tabulated and graphed.

Shifting the record series to create a lag

This particular step in the analysis allows us to shift series to find the lag at which their correlation is maximal

```
Lag = 0
Series A:  5 3 7 6  1  2 4 8 ...
Series B:  9 7 5  4 2  1 0 1 ...

Lag = 1
Series A:  -   5 3  7 6  1 2 4 8 ...
Series B:  9   7 5  4 2  1 0 1 ...


Lag 2
Series A:  - -   5 3 7 6 1 2 4 8 ...
Series B:  9 7   5 4 2 1 0 1 ...
etc.
```

Eigenmaps

An Eigenmap is an efficient map, also called a characteristic map. It is a discontinuous (sample) image of the true continuous trace of the trajectory. Because of this, the connecting line segments in the map are imaginary constructs corresponding to nothing tangible in reality, other than the straight line distance of the connected points.

The trajectory points describe real objects or the states of a real process. Each mapped point is in fact the tip of a paleorelevé vector in phasespace. Because of this, the Eigenmap is a map of the palynological spectrum in linear phase space.

The definition of process states by paleorelevés is the sole realistic basis for the measurement of the compositional transition process in palynological spectra. On the count of this, the contemplative observer is left with more than a lingering thought of how easily scale selection can affect the placement of a trajectory point into the Eigenmap and define what he or she will see of the real process.

We perform Eigenmapping following equal time-step transformation in anticipation of subsequent steps in the analysis that require such transformation. The Eigenmaps of SH, BT, and TDB are in the stereo-grams of Fig. 3 constructed after equal time step transformation uniformly to equal 50 ^{14}C years. .

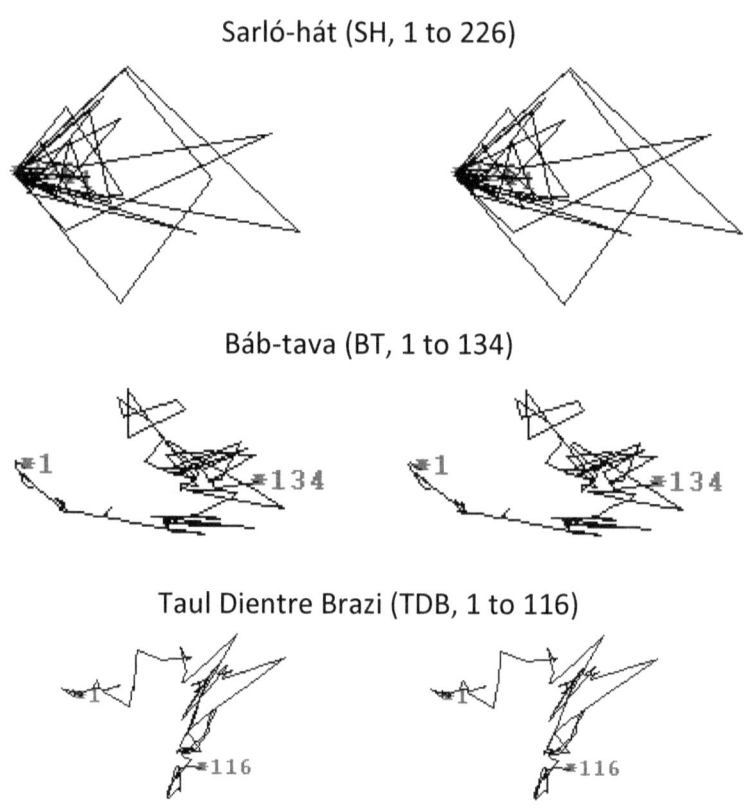

Fig. 3 Stereo images of the efficient trajectory from sites SH, BT and TDB. The reference system's axes are Eigenvectors (replacing the palynomorph taxa). The period lengths are *226, 134,* and *116* in *50 ^{14}C yr* time steps, numbered from present (1) to past. Time span in years in same order as the sites: *80 to 11380 BP, 1138 to 7800 BP,* and *9956 to 15755 BP*. The distance configuration of points within the 3-dimensional stereograms account for *81.0%, 89.9%* and *90.7%* of the original, full-dimensional distance configurations of trajectory points. The trajectory line is

the time axes of reference space. The stereograms should rise into three dimensions for the practiced eyes without being aided by a stereoscope, and certainly with the aid of a stereoscope when the left and right stereo-grams are offset by enlargement or reduction to about 6 cm between identical points.

Obviously, the stereo maps are highly efficient. After the time steps having been adjusted to equal width, the distance of any two points in the stereograms is a close approximation to the true average velocity of compositional transitions within the time interval between the points. The grater the difference between adjacent segments, the greater is the compositional transitions acceleration or deceleration (symbolized by letter A). Generally:

$$A_{12} = \frac{V_2 - V_1}{|t_1 - t_2|}$$

In the equation, V_1 and V_2 represent the average transition velocity within the segments; t_1 and t_2 are the time points measured from initiation (point **A** in Fig. 2) at the segments end points.

A positive A_{12} indicates acceleration and a negative A_{12} indicates deceleration. Saying this in yet another way, given L1 and L2, the lengths of two successive trajectory segments of equal time units, acceleration occurs when L2 > L1. We refer to the historic period of explosive A oscillations as a hotspot of compositional transitions.

Any trajectory point, except the first and the last, has a lead-in line segment and a follow-up line segment. The acute angle enclosed by two line segments is indicative of the intensity of qualitative transition. Qualitative implies proportionality in taxon representation, entrance of new taxa into the records, or the dropping out of taxa from the records. A general rule: the sharper the inner angle, the more dramatic is the qualitative transition.

Multiscale correlation

Two time series are considered. One is the compositional acceleration (deceleration) values series *A* and the other the Vostok temperature series *T*. The *A* series can be distance or angular based. In the present case it is distance based. When should we use an angular *A*? -- if the qualitative aspects of compositional transitions were to be emphasised.

We mentioned already, the *T* series contains temperature differences. The temperature standard is based on the current deuterium content of the oceans. The immediate objective of multiscale correlation analysis is to verify if the *A* and *T* series are statistical linked. When a significant linkage is found, the next step is to determine the time lag at which the linkage is strongest.

The *A* graph of Sarló-hát

We present the Sarló-hát *A* series juxtaposed with the Vostok temperature differences graph *T* in Fig. 4. The *A* graph as given is distance based.

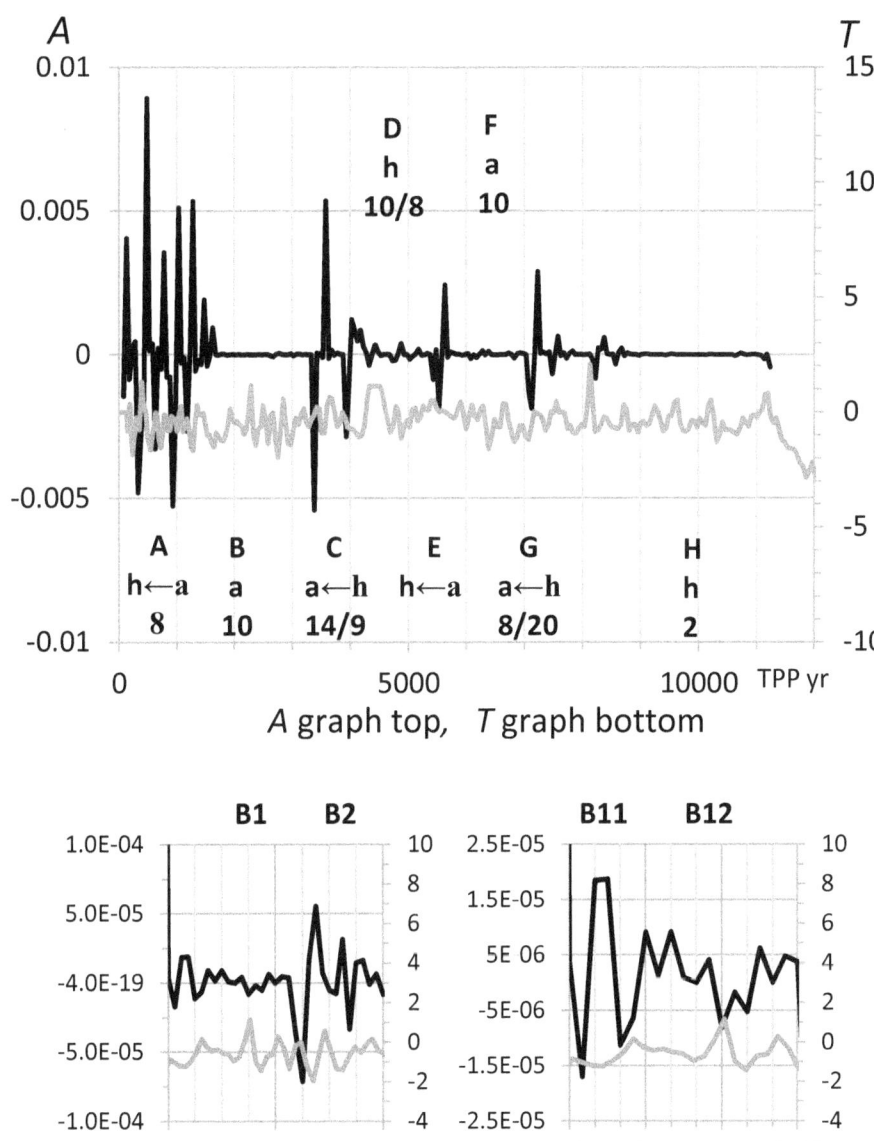

Fig. 4. The *A* and *T* graphs on top span the total SH period. The sections **H** to **A** designate characteristic periods in the Holocene Epoch. Letter h indicated more humid climate. Letter "a" indicates a more arid climate. Numerals 8 to 2 identify the Lag at which the *A,T* series have maximum correlation. *S*ection **B** (two bottom graphs) is selected to show examples

of low-level structures which emerge into visibility when the resolution of the *A* scale is high. Major hotspots on the *A* graph: **A, C, E, G**. Minor hotspots within period **B**: **B**2, **B**11. Similar structures are found within other periods as well. Further explanations are in the text.

We do not show the details, but multiscale embedding does in fact exists through the entire Holocene Epoch. What to make of such a complex phase structure? It is unlikely that we deal with some unusual system behaviour. It is more likely that Nature works this way within systems in which strong environmental and biological determinism convolute with much randomness.

Time delayed climatic effects

Multiscale correlation analysis performed

The objective is to determine the time lag at which the statistical linkage of the *A* and *T* graphs is strongest. Two forms of the standardised product moment, $r(A,T)$ and $\rho(A,T)$, are centre pieces in the analysis. We refer to the generic type as a product moment correlation coefficient. We apply the $r(A,T)$ and $\rho(A,T)$ in a multiscale context. It is important to note that $r(A,T)$ and $\rho(A,T)$ are linear scalars (Orlóci 2010). We expand on this choice and give justifications in context in the Overview section.

Our $r(A,T)$ is a scale dependent estimators. The $\rho(A,T)$ is defined across scales and derived from the scaled $r(A,T)$ values by regression estima-tion. Although most aspects of the methodology are discussed else-where in detail (Orlóci 2009 and references therein), we still need to clar-ify the precise difference between $r(A,T)$ and $\rho(A,T)$.

A new $r(A,T)$ value is computed at each block size $BS=1,2,3, …$[16] for each value of the lag $L=1, 2, 3, …$. All $r(A,T)$ values are computed for the full

---

[16] It was already mentioned that blocks are moving windows which we use for averag-ing and therefore smoothing the series. In this way, we accomplish a dampening of

workable length of the smoothed A and T series. The smoothed series contains averages calculated for each position of the BS sized moving window (see example earlier in the text). For the Sarló-hát A series, taken at full length, we have the $r(A,T)$ values corresponding to $L=0$ listed in Table 1.

Table 1. Sarló-hát statistics for the A,T comparisons. Legend to symbols: BS block size (moving window size within which the elements are averaged at each position); $r(A,T)$ product moment correlation; F^+%, F^-% per cent of the positive and negative correlation values based on random sampling* of the A,T pairs in the entire series a very large number of times (further comments in the text); $\rho(A,T)$, Φ^- and Φ^+ solutions of the linear regression equation for $BS=1$ based on $r(A,T)$, F^+% or F^-% as the y variable, on BS as he x variable at $L=0$ (see details following the table). Full series length: 224. Note the discrepancy percentages: $100 - 42.5 - 18.8 = 38.7$. This is the proportionate number of 'zero' correlation values (within computer rounding errors) found in sampling and resampling of the A,T series at block size 1.

L	BS	r(A,T)	F⁻ %	F⁺ %
0	1	-0.030	42.5	18.8
0	2	-0.032	48.8	27.0
0	3	-0.027	45.6	22.1
0	4	-0.024	42.3	23.2
0	5	-0.034	47.7	22.6

$$\rho(A,T) = -0.029 \quad \Phi^- = 44.6 \quad \Phi^+ = 22.0$$

*The random sampling and resampling experiment involves taking a large number of samples, each involving at least 5 A,T pairs randomly from the entire A and T series. The upper limit in the example is 40 A,T pairs per sample. Statistical considerations of error control apply. Each random sample mentioned supplies its own $r(A,T)$ value which we score

---

random variation. The upper limit of BS depends on the length of the A series $r(A,T)$ and $\rho(A,T)$.

for its + or - sign. The scores are counted. The counts are the basis of the F^-% and F^+% percentages.

Regression analysis concludes this set of analytical steps. Fig. 5 contains the characteristic graphs and regression equations based on the data in Table 1. The solutions of the regression equation for *BS=1* hand as the $\rho(A,T)$, Φ^- and Φ^+ estimates.

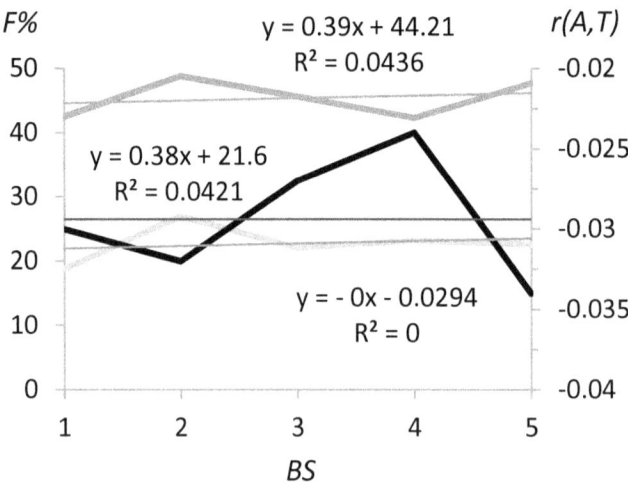

Fig. 5. Graphs based on data in Table 1. Sequence of graphs on extreme right: F^-, F^+, $r(A,T)$. Vertical axes: left F^-%, F^+%; right $r(A,T)$. Horizontal axis: block size *BS*. Full series (period) length on which the calculations are based: 224 *A,T* pairs. Dominant correlation at *L=0*: negative. Regression equations of straight lines:

$$\rho(A,T) = 0BS - 0.0294 \quad \Phi^- = 0.39BS + 44.21 \quad \Phi^+ = 0.38BS + 21.60$$

In the regression equations y is $\rho(A,T)$, Φ^- or Φ^+, x is *BS*. See regression estimates at *BS = 1* in Table 1. R^2 is the coefficient of determination (see Orlóci 2010 for regression terminology).

After having calculated $\rho(A,T)$, Φ^-, and Φ^+ at *L=0*, we repeat the same analysis again at each increment in the lag variable *L=1, 2 ... u*. Letter *u* stands for an arbitrary upper bound. Finally, we end up with *u+1* sets of $\rho(A,T)$, Φ^- and Φ^+ estimates. We present the results for different time periods in the individual plates of Fig. 6. At any value *L*, the time lag is *L* x 50 ^{14}C years

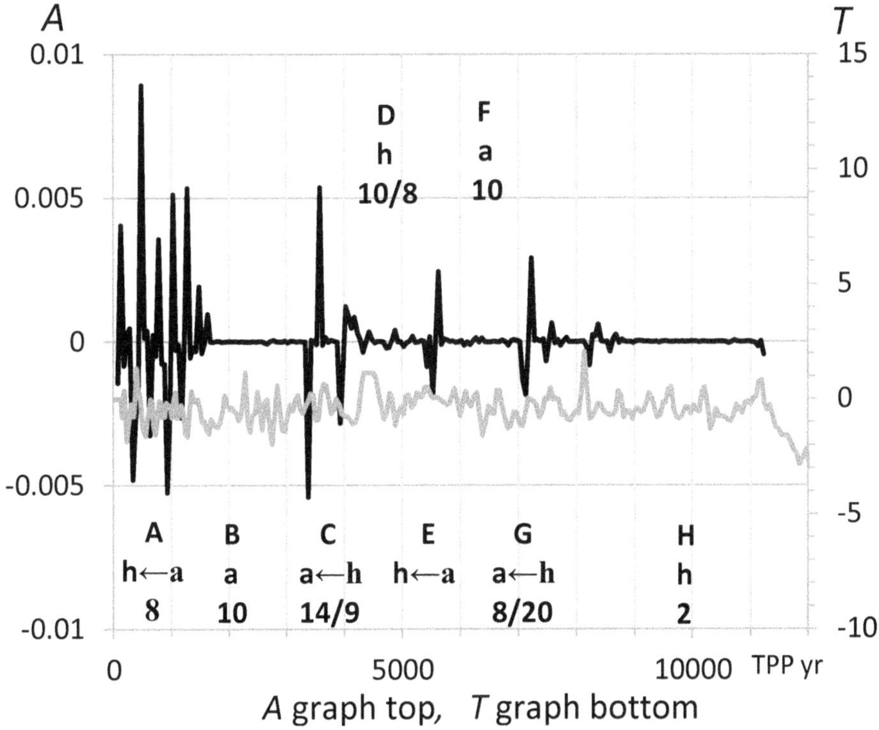

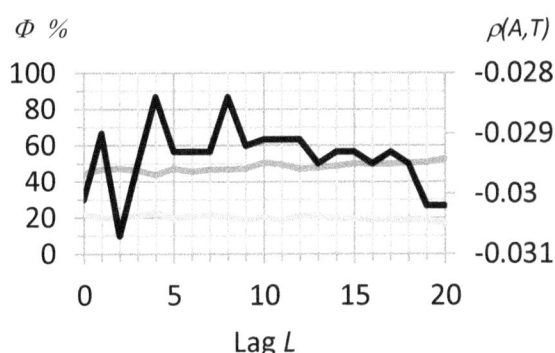

A graph top, T graph bottom

Full period: 81- 11231 ^{14}C yr; 224 *A,T* pairs. Sequence of graphs at extreme right: Φ^-, $\rho(A,T)$, Φ^+

Lag *L*

Period **H**: 8631-11231 ^{14}C yr; 53 A,T pairs. Sequence of graphs at extreme right: Φ^+, $\rho(A,T)$, Φ^-

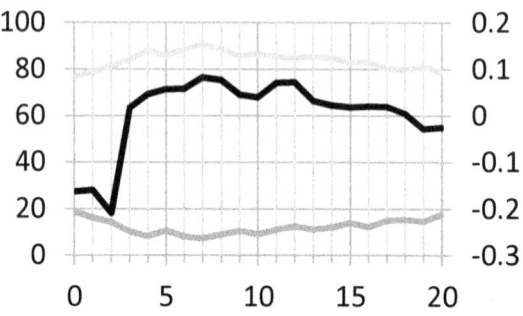

Period **G**: 6931- 8631 ^{14}C yr; 35 A,T pairs. Sequence of graphs at extreme right: Φ^+, $\rho(A,T)$, Φ^-

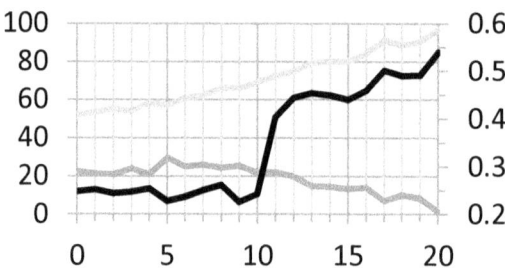

Period **F**: 5731-6931 ^{14}C yr; 25 A,T pairs. Sequence of graphs at extreme right: Φ^-, $\rho(A,T)$, Φ^+

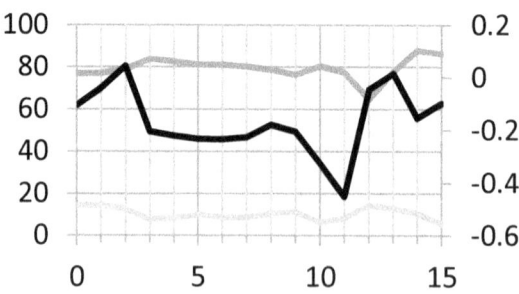

Period **E**: 5331-5731 ^{14}C yr; 9 A,T pairs.
(Period too short for analysis)

Period **D**: 4431-5331 ^{14}C yr; 19 *A,T* pairs. Sequence of graphs at extreme right: Φ^+, $\rho(A,T)$, Φ^-

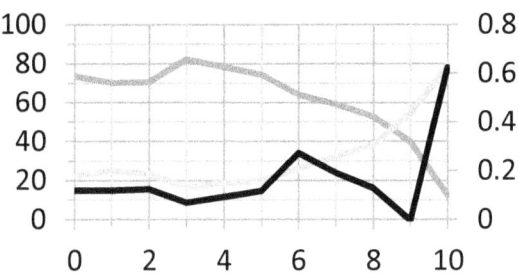

Period **C**: 3281-4431^{14}C yr; 9 *A,T* pairs. Sequence of graphs at extreme right: Φ^-, Φ^+, $\rho(A,T)$

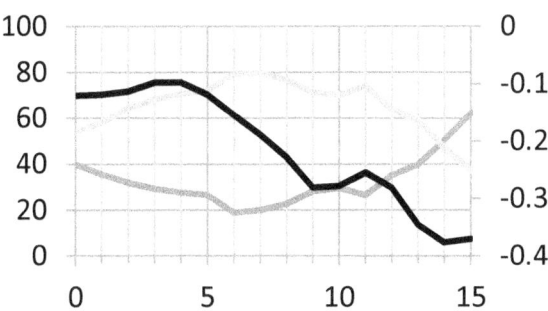

Period **B**: 1681-3281^{14}C yr; 33 *A,T* pairs. Sequence of graphs at extreme right: Φ^-, $\rho(A,T)$, Φ^+

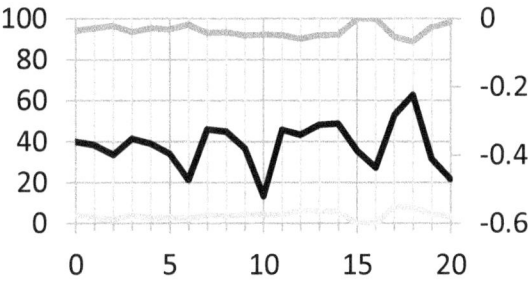

Period **A**: *81*-1681^{14}C yr; 33 *A,T* pairs. Sequence of graphs at extreme right: Φ^-, $\rho(A,T)$, Φ^+

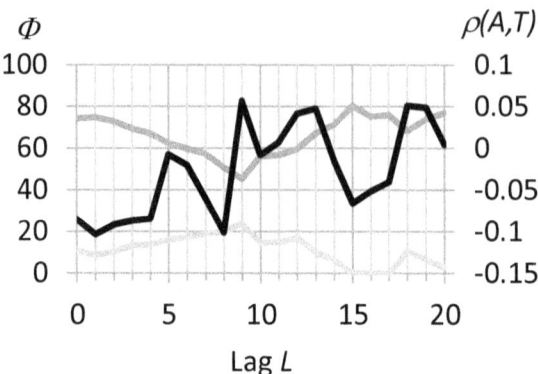

Fig. 6. First plate A,T graphs for full SH period copied from Fig. 4. The individual plates contain the $\rho(A,T,\, \Phi^-,$ and Φ^+ graphs for time periods **H** to **A**. The time period, number of A,T pairs, and sequence of the graphs vertically are identified in the plates' caption. Upper limit of the lag L varies depending on the size of the time period.

We can address substantively two fundamental question based on the $\rho(A,T), \Phi^-,$ and Φ^+ graphs:

Are the A and T series synchronous?
If they are, at which lag is synchrony maximal?

Before we provide the answers we have to tell the reader how to interpret the graphs.

Interpretation of the $\rho(A,T)$ graphs

There are specific points to be made at the outset to create context for interpretation of the $\rho(A,T)$ values:

i. Any observed $\rho(A,T)$ value is a statistic. As such it has a probability associated with it on the basis of which we can judge its rarity as a chance event. More explicitly, the probability will tell how likely it is to obtain a correlation value at least as extreme as the observed $\rho(A,T)$ value when the compositional transitions in the palynomorph assembly are ruled by chance and therefore the expected correlation is zero. The smaller the probability, the more attractive is the alternative interpretation that $\rho(A,T)$ is not a freak event, but a significant, trustworthy value whose expectation is not zero.

ii. How do we determine the probability associated with an observed $\rho(A,T)$ value? It would not be wise to use the usual statistical tables. The reason is simple: those tables were generated under strict regularity conditions (Orlóci 2010), including random sampling within a homogeneous statistical universe defined by the bivariate normally distribution of the A,T series. These are to confining. But we do not need to go that way. Probabilities can be found in a different way when we use Monte Carlo type experiments. In this the only thing to assume is the random nature of the process that rules the palynomorph composition in the spectrum. Naturally the rule of chance implies a zero correlation. Under this condition we will find that the expectation of $\rho(A,T)$ is zero in the long run.

The Monte Carlo experiment which serves our need includes randomization in one of the series, say A, and sampling and re-sampling of the A,T duplets in the randomized series a very large number of times. The results are a set of probability points ρ for $\rho(A,T)$ and probabilities P associated with ρ. The Monte Carlo experiment produced the results for Sarló-hát which we present by graph in Fig. 7.

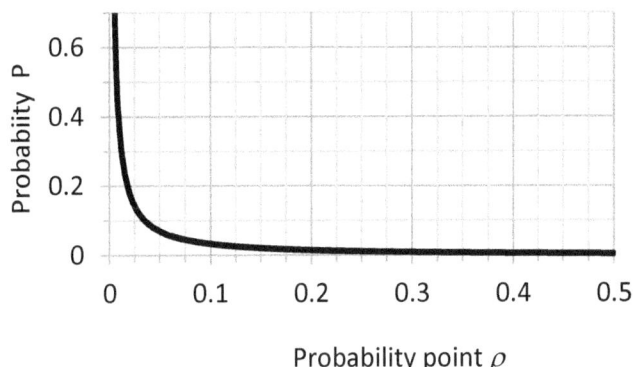

Fig. 7. Probability distribution graph generated in Monte Carlo experiment under zero expectation for $\rho(A,T)$. *P* is the probability of a randomly selected $\rho(A,T)$ being at least as extreme as any given probability
point ρ under the assumption of zero expectation. See the main text for
important details.

The curve of the *P* values in Fig. 7 has functional form

$$P = \frac{\rho^{-1} - 1.181207}{267.93106}$$

This is a probability distribution function referring to all cases at least as
large as ρ. How do we read the graph? Wen *P* associated with a value of
$\rho(A,T)$ is small, say 0.05 or smaller, we may reason that such a $\rho(A,T)$ is
unlikely to have zero expectation and we declare the value significant.
The probability that we are wrong in doing so, is *P*.

Example:

Observed $\rho(A,T) = 0.068$
Monte Carlo probability of an at least as large correlation value as 0.068
is

$$P = \frac{\rho^{-1} - c}{b} = \frac{\frac{1}{0.0686} - 1.181207}{267.93106} = 0.0499$$

With 0.05 as the threshold value, we have to regard *0.0499* significant.
The probability that we made an erroneous decision with declaring 0.068
significant is 0.0499. We note, under the actual 10000 iterations in the
experiment which generated the probability distribution even a numerically weak $\rho(A,T)$, such as in the example, can attain statistical significance.

Derivation of probabilities in a Monte Carlo experiment has consequences:

i. High local relevance owing to the probabilities being based on the observations we are analysing.

ii. Low generality across cases. The ρ, P graph has no utility to test $\rho(A,T)$ values from external data sets.

iii. A significant $\rho(A,T)$ can be numerically strong or weak. A significant and numerically strong $\rho(A,T)$ gives us confidence to assert that temperature oscillations *(T)* go in tandem with the compositional transition's acceleration *(A)*. This can mean that *T* has a direct effect on *A* or *T* is proxy for some other temperature related forcing.

iv. Ambiguities will be faced when $\rho(A,T)$ is found numerically trivial. In that case, *A* type oscillations are either independent from local temperature related forcing or the Vostok temperature oscillations are not translatable into effects sufficiently local to influence the oscillations of *A*.

Interpretation of the Φ^- and Φ^+ graphs

Do we have any process governance rule to fall back upon when we interpret the $[\Phi^- \ \Phi^+]$ frequency distribution? Yes, we have. In fact Orlóci (2009) and Orlóci and He (2009) has shown, taking an observed regularity extracted from 17 palynological spectra from sites spread across contrasting climatic zones from tundra to hot desert across the Globe, that the dominance of Φ^- is indicative of climatic aridity, but the dominance of Φ^+ is indicative of climatic humidity. This a conclusion was deduced from the almost function negative correlation of Φ^- and similarly the almost functional positive correlation of Φ^+ with the Thornthwaite index *(Th)*.

There is more to their conclusions:

i. Considering that Φ^+ is the frequency of positive correlation of compositional transition acceleration and atmospheric temperature, and considering further that Φ^+ is positive correlated with atmospheric humidity, the dominance of Φ^+ must indicate that under humid atmospheric conditions increasing atmospheric temperature will accelerate compositional transitions in the palynomorph spectrum.

ii Considering that Φ^- is the frequency of negative correlation of compositional transition acceleration and atmospheric temperature, and considering further that Φ^- is negatively correlated with atmospheric humidity, the dominance of Φ^- must indicate that under arid atmospheric conditions increasing atmospheric temperature will decelerate compositional transitions in the palynomorph spectrum.

The data* referred to above is now reproduced and subjected to a complete regression analysis by a linear model:

#	Th	Φ^+	Φ^-
1	1.43	80.96	16.3
2	1.88	80.9	16.76
3	1.25	75.84	20.68
4	1.88	68.68	30.22
5	1.88	65.08	29.16
6	1.25	57.31	38.39
7	1.75	56.63	32.77
8	0.54	52.79	41.25
9	0.71	46.84	39.99
10	0.71	42.62	55.08
11	0.36	33.64	63.71
12	0.63	15.12	84.32
13	0.31	11.15	84.91
14	0.31	4.74	93.91
15	1.39	65	29.5
16	0.46	21.45	76.39
17	1.04	49.45	46.25

*See Table 5 in Orlóci 2009.

In the above: $r(Th, \Phi^+)=0.833$ and $r(Th, \Phi^-)= -0.823$.

Regression analysis Th on Φ^+

	Value	Std err	t-value	95% Conf lim		$P(t_{RND} > |t|)$
Intercept c	0.064916	0.18687	0.347384	-0.33339	0.463221	0.73313
Reg coeff b	0.020136	0.003455	5.827871	0.012771	0.0275	0.00003

Source	SS	DF	MS	F	$P(F_{RND} \geq F)$
Regr	3.803724	1	3.803724	33.9641	**0.00003**
Error	1.679888	15	0.111993		
Total	5.483612	16			

R^2	DF adj R^2	Fit STD ER	F-value
0.693653	0.649889	0.334653	33.96408

Regression analysis of Th on Φ^-

	Value	Std err	t-value	95% Conf lim		$P(t_{RND} > |t|)$
Intercept c	1.949389	0.18159	10.73509	1.562338	2.336439	0
Reg coeff b	-0.01921	0.003429	-5.60123	-0.02652	-0.0119	0.00005

Source	SS	DF	MS	F	$P(F_{RND} > F)$
Regr	3.70989	1	3.70989	31.3738	**0.00005**
Error	1.773722	15	0.118248		
Total	5.483612	16			

R^2	DF adj R^2	Fit STD ER	F-value
0.676541	0.630333	0.343872	31.37377

Legend to symbols: SS - sum of squares; DF - degrees of freedom; MS - mean square; F - variance ratio statistics; $P(F_{RND} \geq F)$ - probability of an at least as large random F as the observed F under the rule of chance; R^2 - coefficient of determination; Fit STD ER - standard error of the fit; adj - adjusted; Conf lim - confidence limits. Find definitions in Orlóci (2010) and on the internet. It is quite clear from the parameter values in bold that the $Th \times F$ relationship is strong and very highly significant.

The sample graphs to be interpreted are in Fig. 8. The numerical values are in Table 2. The horizontal axis represents the lag variable L. The algorithm allows any lag to be used, but the choice must be reasoned. When we put $L=0$, the A and T series are not displaced. At $L=x$, the elements in the *A* series are paired with temperature values which occurred 50x tears earlier in the past. For example, *L=20* defines a lag of 1000 years.

Table 2. $\rho(A,T)$, Φ^- or Φ^+ are estimates for time period H (*8631 – 11231 BP*). Regression equations fitted to the data are given in Fig. 8.

L	$\rho(A,T)$	Φ^-	Φ^+	L	$\rho(A,T)$	Φ^-	Φ^+
		%	%			%	%
0	-0.163	18.84	76.64	11	0.0709	11.22	85.50
1	-0.1599	16.22	78.70	12	0.0725	12.58	84.76
2	-0.2089	14.32	81.44	13	0.0326	11.26	85.52
3	0.0171	10.38	84.16	14	0.0227	12.14	84.78
4	0.0460	8.34	88.30	15	0.0189	14.06	82.96
5	0.0576	10.72	85.82	16	0.0209	12.439	83.22
6	0.0581	8.30	88.86	17	0.0202	15.119	80.46
7	0.0827	7.8	90.68	18	0.0048	15.62	79.68
8	0.0774	9.06	88.80	19	-0.0281	14.78	81.06
9	0.0455	10.58	85.56	20	-0.0255	17.72	78.76
10	0.0399	9.22	86.62				

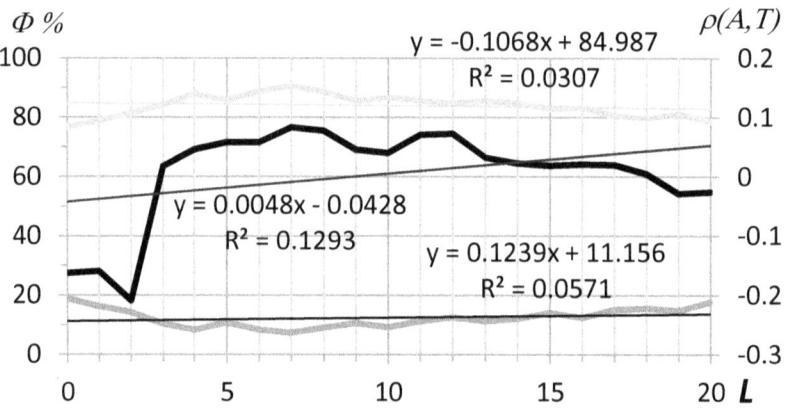

Fig. 8. Graph drawn based on the records in Table 2. The time period covered is section **H** from 8631 to 11231 BP (dates we read directly from the graph but not consider to be exact).. Straight lines correspond to re-gression equations fitted to the graphs. In regression equations y stands for $\rho(A,T)$ (middle), Φ^- (bottom), Φ^+ (top); x stands for L. R^2 is the coef-ficient of determination. The inverse tangent of -0.1068 is $6.1°$ for Φ^+. The coefficient's negative sign indicates descent. The Φ^- graph shows

ascent on a somewhat steeper slope 7.1°. The $\rho(A,T)$ graph's average assent is 0.3°. [17]

At each lag setting, the analysis hands us a single characteristic value for $\rho(A,T)$, Φ^- and Φ^+. If we were to use a single triplet $\rho(A,T)$, Φ^-, Φ^+ to characterize the effect of lag within the time period under consideration, we will make it a convention to choose the maximum $\rho(A,T)$ and take with it the corresponding Φ^- and Φ^+ values. The maximum $\rho(A,T)$ occurs at $L=2$ or *100* years in Table 2.

What do these mean?

We asked two short questions earlier about the delayed effect and now give a long answer. In the course of this we shall interpret the Φ^- and Φ^+ graphs.

What is a striking feature of the graphs in Fig. 8 and the numerics in Table 2, is the narrow range of $\rho(A,T)$ and the wide separation of the Φ^- and Φ^+ graphs. The regression slopes are not steep. The maximum absolute value of $\rho(A,T)$ is *0.2089* at $L=2$ at which $\Phi^- = 14.32$ and $\Phi^' = 81.44$. These tell us that the lag effect on $\rho(A,T)$ at the maximum is six folds the regression expectation *0.0332* . Where do we get the *0.0332* value from? It is $|0.0048 \times 2 - 0.0428| = 0.0332$ (see equation in Fig. 8). The ratio is *0.2089/0.0332 = 6.29*.

What should we conclude from these for time period **H**? For one thing we take the absolute *0.2089* value highly unusual. When referred to the

---

55

graph in Fig. 7, the probability of a correlation values at least as large as *0.2098* under the zero expectation rule is

$$P = \frac{\rho^{-1} - c}{b} = \frac{\dfrac{1}{0.2098} - 1.181207}{267.93106} = 0.013381$$

We consider 0.2098 an outlier, highly significant correlation. We conclude the following:

i. The historic temperature effect is strongest in the retrospect of *2 x 50 = 100* years.

ii. The absolute dominance of Φ^+ indicates humid climate for period **H**.

A similar method of interpretation applies to all segments. We summarize the details in Table 3.

Table 3. Observed $\rho(A,T)$ maxima and associated Φ^- and Φ^+ estimates for time periods **A** to **H** at Sarló-hát for *BS=1*. Letters "a" and "h" in last row: more arid, more humid; BP: time before present in ^{14}C yr units; *L*: time lag in 50 year units at which the value of $\rho(A,T)$ reach maximum within the period; *b*: regression coefficient for $\rho(A,T)$ over *BS* (copied from computer output corresponding to graphs in Fig. 6); *c*: value of the regression line intercept on the vertical axis; *y*: regression estimate for $\rho(A,T)$; R^2: coefficient of determination; *Th* on Φ^- and *Th* on Φ^+: regression estimates of *Th* based on Φ^- and Φ^+ computed by the regression equations given in Fig. 7: $\leftarrow$: climate type transition.

Period	A	B	Cb	Ca	Db
TBP	81 - 1681	1681 - 3281	3281 - 4431	3281 - 4431	4431 - 5331
L	8	10	14	9	10
$\rho(A,T)$	-0.101	-0.52	-0.37	-0.281	0.624
b	0.006	0.001	-0.02	-0.02	0.022
c	-0.082	-0.38	-0.064	-0.064	0.057
y	-0.038	-0.37	-0.344	-0.244	0.281
R^2	0.377	0.008	0.901	0.901	0.201
Arc tan ^{o}	0.315	0.057	-1.146	-1.146	1.283
Φ^-	50.900	92.340	62.280	28.180	12.420

Φ^+	19.740	4.300	37.020	71.340	78.980
Th on Φ^-	0.972	0.176	0.753	1.408	1.711
Th on Φ^+	0.462	0.152	0.810	1.501	1.655
Climate type	h←a	a	a	a←h	h

Period	Da	F	Ga	Gb	H
TBP	4431 - 5331	5731 - 6931	6931 - 8631	6931 - 8631	8631 - 11231
L	8	10	8	20	2
$p(A,T)$	0.129	-0.452	0.226	0.541	0.209
b	0.022	-0.003	0.017	0.017	0.005
c	0.057	-0.143	0.18	0.18	-0.043
y	0.236	-0.175	0.316	0.316	-0.033
R^2	0.201	0.014	0.814	0.814	0.129
Arc tan °	1.283	-0.183	0.974	0.974	0.275
Φ^-	52.600	77.380	25,44	1.360	14.310
Φ^+	38.500	8.220	65.900	96.860	81.430
Th on Φ^-	0.939	0.463	1.461	1.923	1.674
Th on Φ^+	0.840	0.230	1.392	2.015	1.705
Climate type	h←a	a	a	a←h	h←a

Tab. 3 excludes period **E** which is too short for reliable analysis. The *L* value given corresponds to maximum absolute $p(A,T)$ within a given time period is the reference. It is an interesting fact that, excepting period **H**, the optimal *L* is at least 400 years. This is the same as saying that the climatic temperature events most highly correlated with vegetation transitions at time *t* occur at least 4 centuries earlier at *t - 400*. The other fact to be noted is the presence of nontrivial correlations at other values of *L* as well, which tell us of the existence of a cumulative effect forward in time. A related topic, the global vegetation events expected under accelerated rates of transition under predicted rates of global warming is discussed by Orlóci (1993, 2008).

The sets of graphs in Fig. 9 capture a very regular trend of changes over the Holocene epoch. Periods of climatic aridity or humidity is functionally linked to the periodicity of *Th* (Fig. 10). From period **H** to **A** the process goes through more arid to more humid periods in 4 complete cycles.

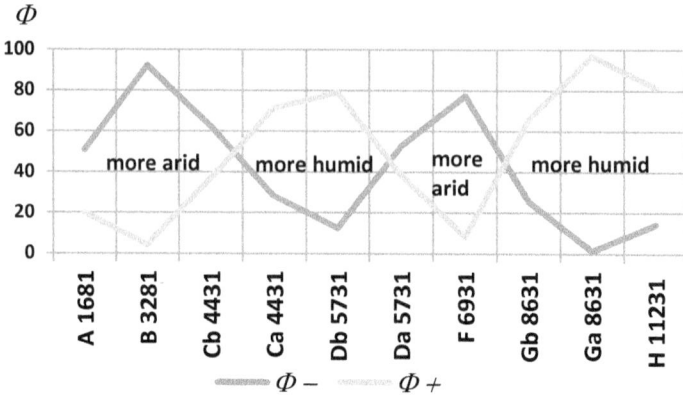

Fig. 9. Long-term historic cycles of Φ^- and Φ^+ during the Holocene Epoch at Sarló-hát. The dominance of Φ^- is characteristic of climates with more tropospheric aridity and the dominance of Φ^+ is characteristic for climates with more tropospheric humidity in terms of the *Th* index (see graph in Fig. 10). The graphs suggest a very regular trend of climate change over the Holocene in Sarló-hát. Symbols: **H** to **A** time periods as in Fig. 6).

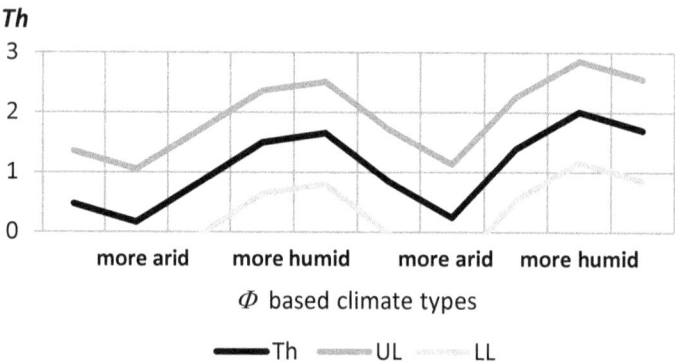

Fig. 10. *Th* periodicity (centre line) and 95% confidence limits *(UL, LL)* on the individual *Th* values. See Table 3 and Fig. 9 for data and functional connection to the Φ parameters.

An interesting phenomenon is revealed by the graphs in Fig. 11. It is concerning the response of the $\rho(A,T)$, Φ^- and Φ^+ parameters to extreme *L* values up to 160 basic time steps (8 k years).

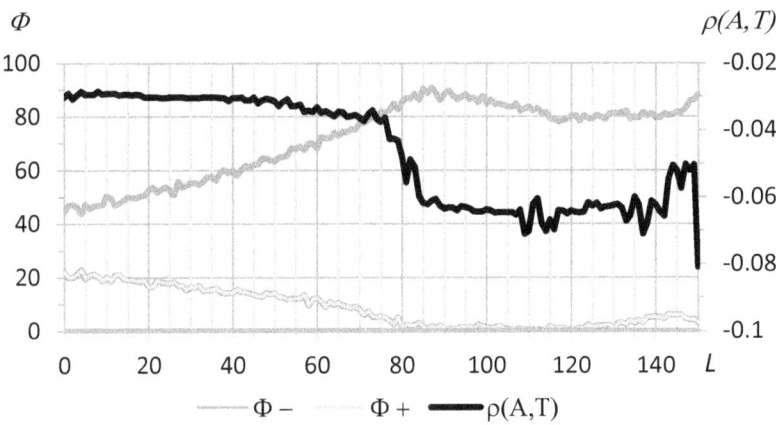

Fig. 11. Graphs of $\rho(A,T)$ (centre heavy line), Φ^- (top) and Φ^+ (bottom) over L up to 160 time steps (8000 year) in Sarló-hát. The graph is based on the full period of 81 to 11231 years.

The effect of the widening L on $\rho(A,T)$ appears minor in absolute terms, but quite dramatic within its actual range, and we should say quite significant in statistical terms (see the probability graph we gave in Fig. 7).

It is quite clear that the greater part of the ρ scale falls within the lower 10 percentile of the probability range. Very low correlation values are included which we judge very unlikely to occur under the rule of chance. For this reason, we should not easily dismiss low observed correlation values as being frivolous.

Observing further the correlation graph in Fig. 11, a striking regularity is evident: up to about $L=75$ or 3750 yr the graph is rather flat, then drops down into a range of statistical significance at better than 0.1 probability. By $L=85$ (4250 yr) the graph flattens but with considerable oscillations starting around $L=110$ (5500 yr), then after approaching the extreme L-$=160$ the graph drops down and the correlation attains statistical significance at better than 0.05 probability.

The effect of increasing L to extreme values is unmistakably colossal by accentuating the difference of the Φ^- (top) and Φ^+ (bottom) graphs.

The ascending slope 19° for Φ^- and the descending slope 11° for Φ^+ up to $L=87$ (4350 years) indicates this. Φ^- graph starts a slight descent at that point up to $L=117$ (5850 years).[18] After that a new flat section begins and then moderate ascent at $L=140$ (7000 yr). The Φ^+ graph mirrors the behaviour of Φ^- in a somewhat dampened manner.

Interestingly, the longer the leg L reaches back to temperature values, the more definite is the imbedded periodic trend, sharpest around $L=95$ (4750 years). This is the approximate wave length of the *Th* index (Fig. 10). We conclude that the compositional transition process do have long physical memory for past climate change.

---

[18] Eq1 $y=-.067x+0.0059$ Eq2 $y=98.529x -7.73$ Eq3 $y=-1.863.64$ up to $L=87$

Indicator taxa

So far we were drawing conclusions from the results of multiscale trajectory analysis about salient parameters of the palynomorph taxa of Sarló-hát. We did not isolate individual taxa out of the assemblage beyond identifying them with the axes of the process trajectory's phase space. We identified periods in the Holocene climate based on examination of the assemblage's total compositional transitions in acceleration terms. We described the periods by generalised parameters and characterized them by their historic temperature and atmospheric conditions. In this section we turn to climate indicator taxa which we select by measurable criteria and describe by their performance as time series in comparable terms.

Previous studies approached the problem of the identification of climate indicator taxa in a very different way. The then current geographic distribution of species was studied and indicator taxa were selected based on the dominant climate under which the taxa existed. Once this was done the distribution of the taxa within the palynomorph spectrum gave clues for the identity of the periodic Holocene climate. There is no reason why this approach should not work, but only up to a point. The multiscale approach, by focussing on the assemblage rather than individual taxa opens much wider vistas for paleoclimate identification in strictly analytical and reproducible ways.

We take up the performance history of selected palynomorph taxa in this section across the entire period length of the Sarló-hát spectrum. For mapping any taxon's performance history, we use *deviation graphs* (Feoli and Orlóci 1985). Deviations are measured from random expectation, a condition highlighted by the rule of chance over the compositional transitions in the palynomorph assembly (see Orlóci 2010).

An observed deviation may be positive or negative. When positive overperformance is indicated, that is the taxon is more common than expected under the rule of chance. A negative deviation indicates underperformance. We display the deviation graphs of 15 taxa in Fig. 12. We also display two *A,T* diagrams the second time for convenient orientation. Limiting the number of the displayed taxa to 25 is a purely space saving decision.

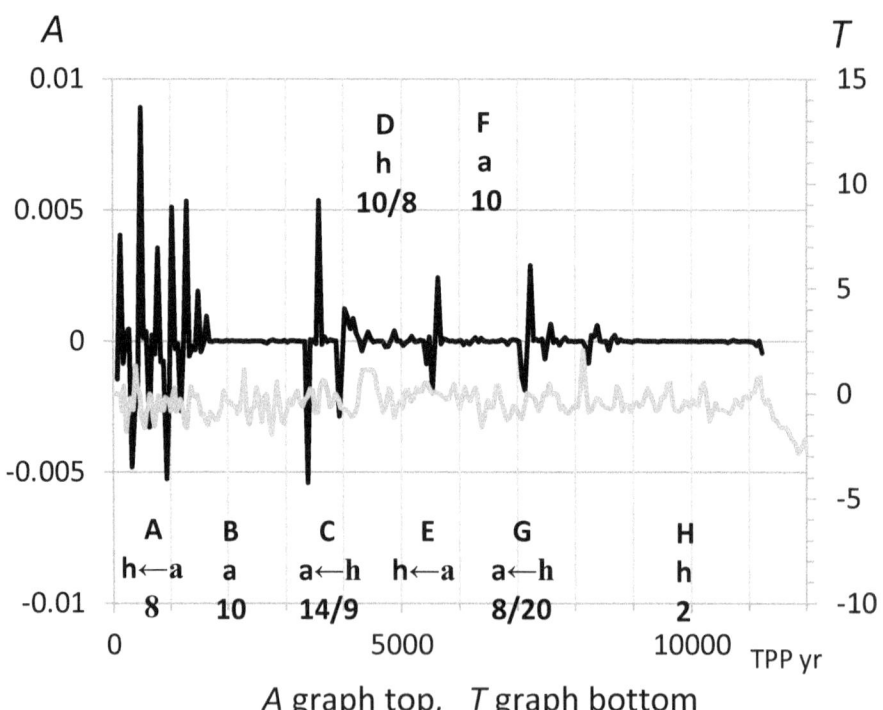

A graph top, *T* graph bottom

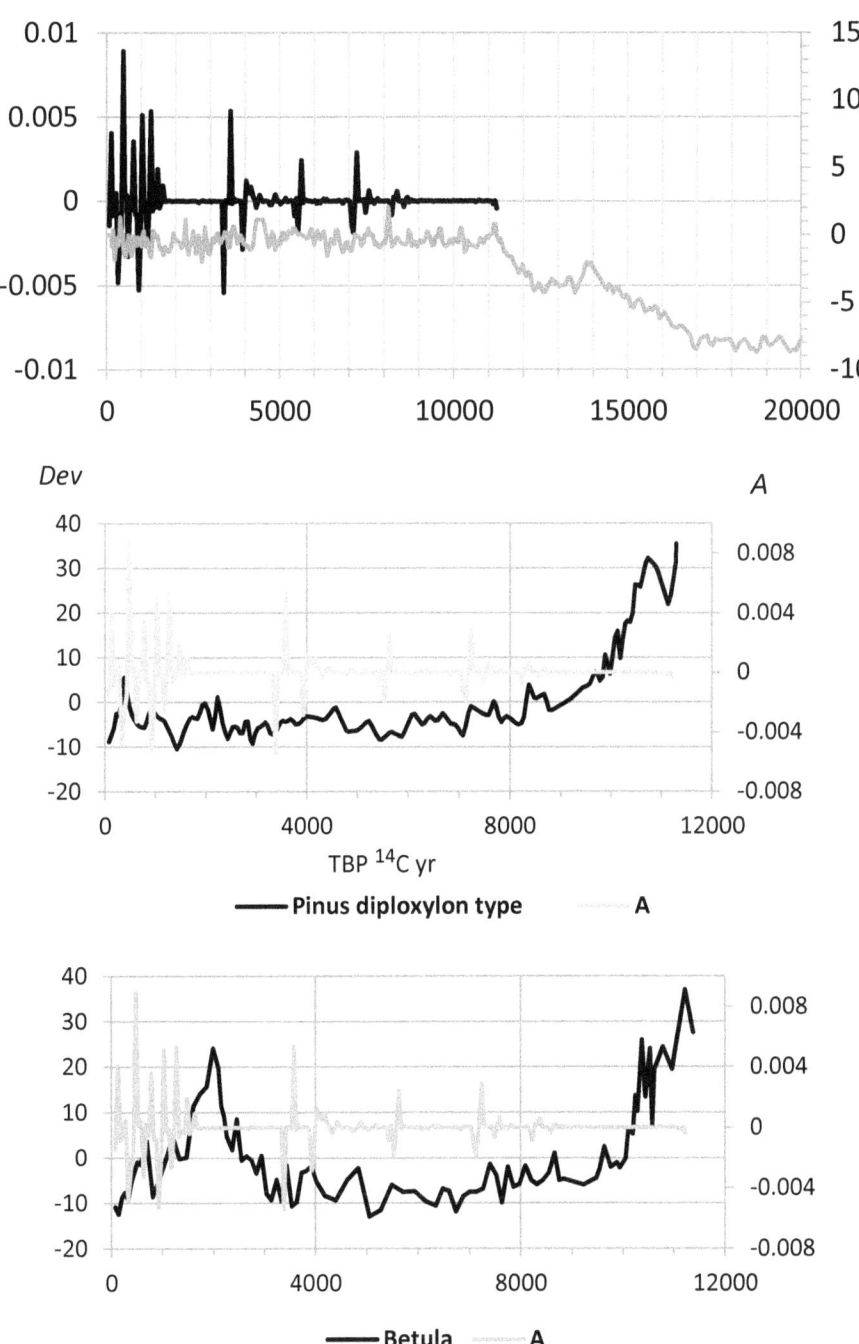

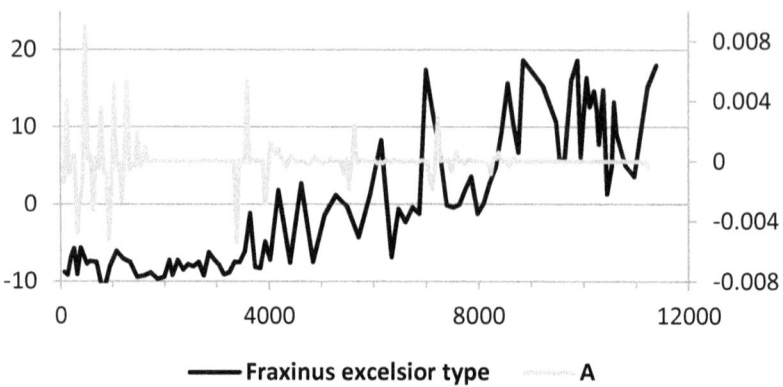

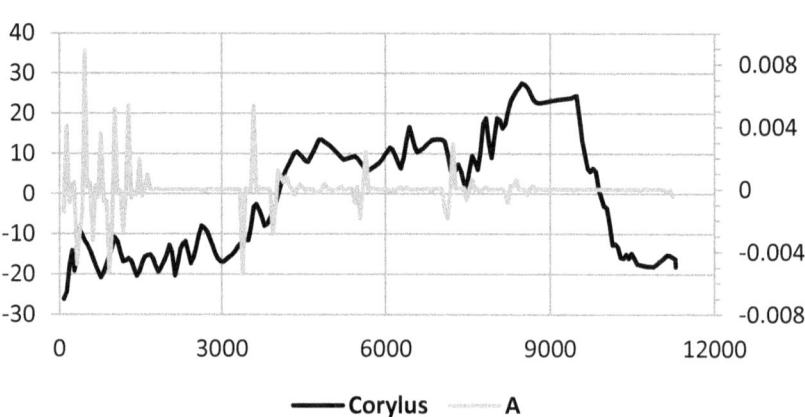

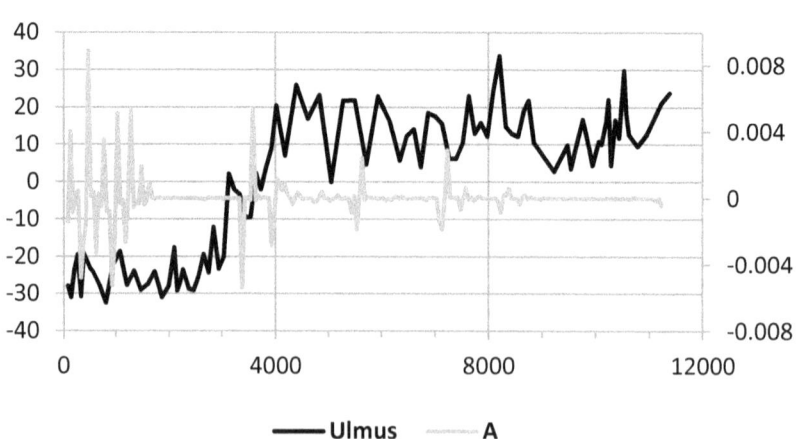

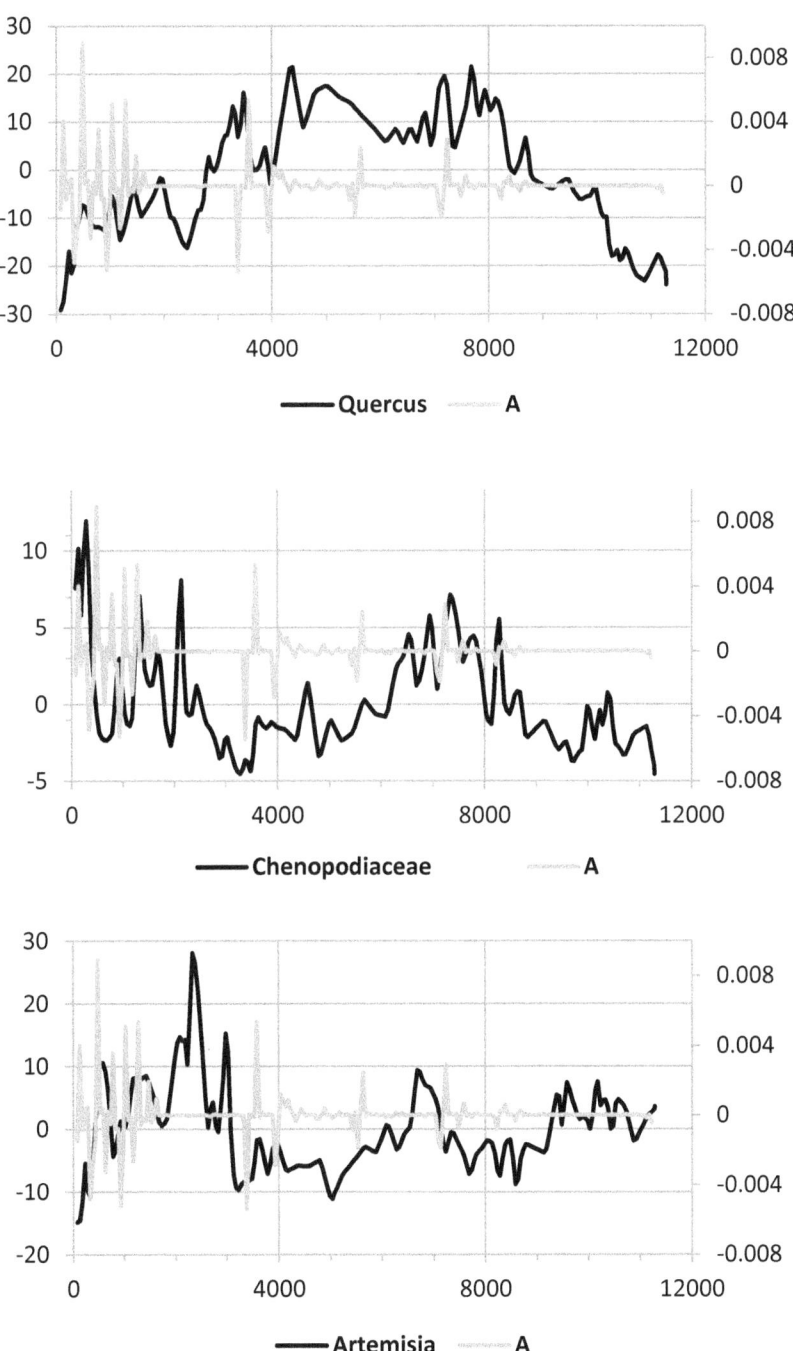

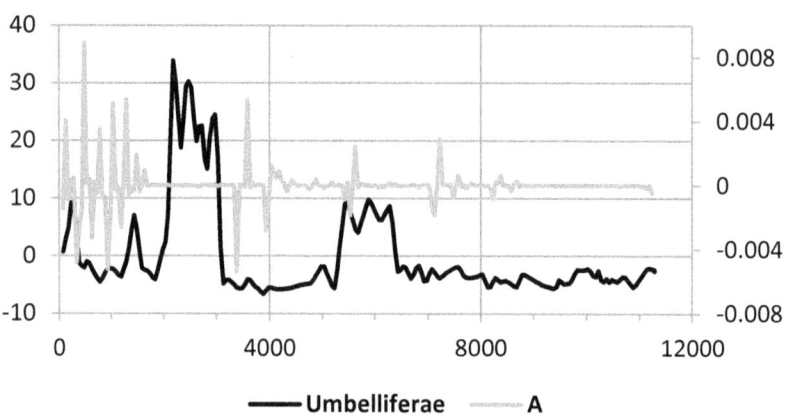

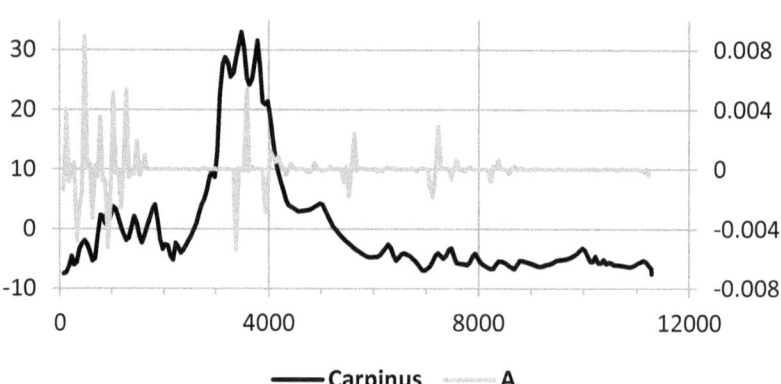

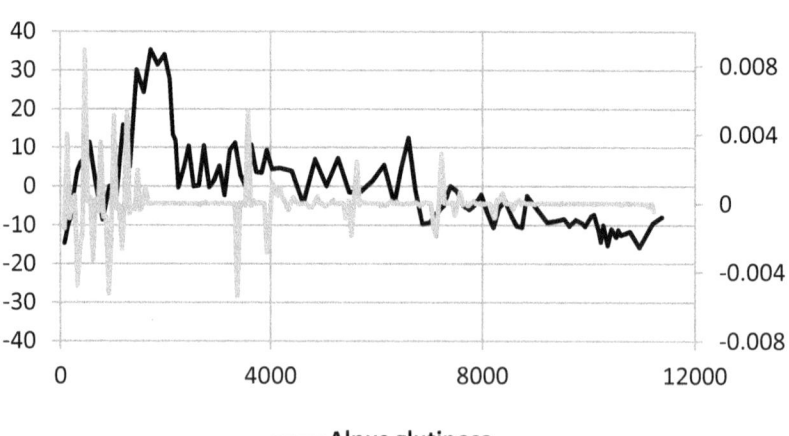

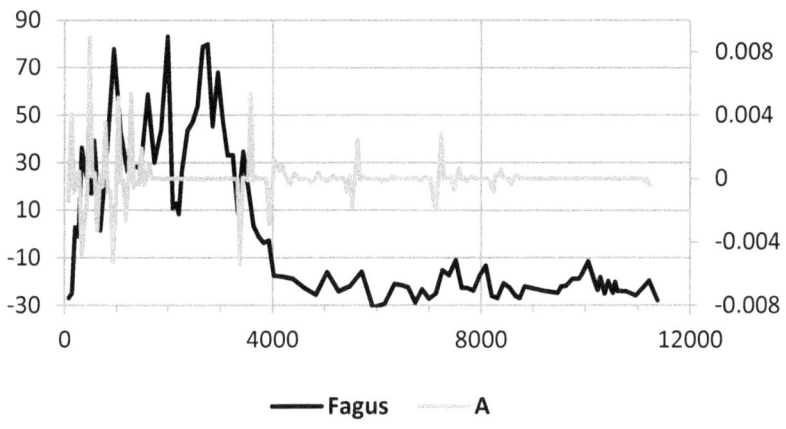

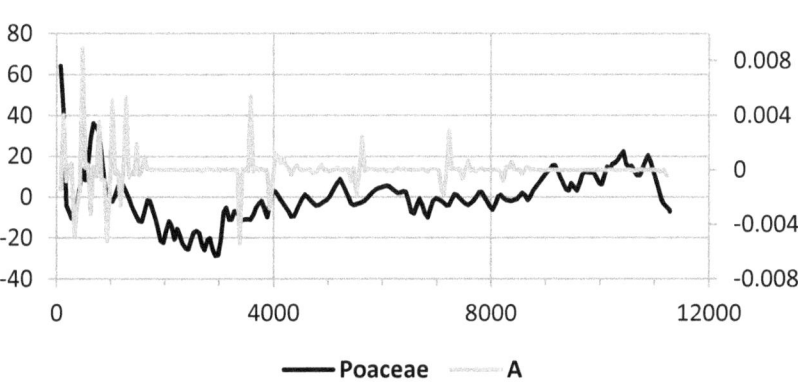

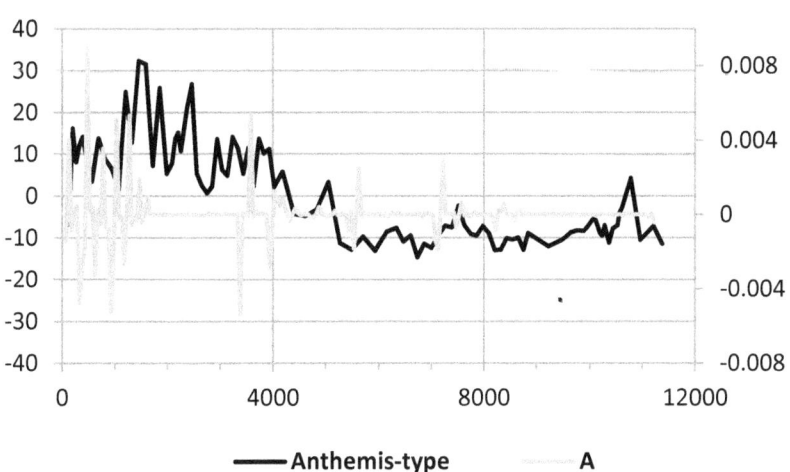

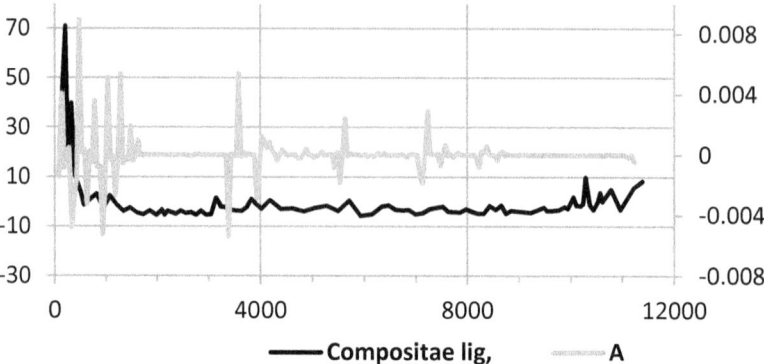

Fig. 12. First two diagrams: *A,T* graphs for full period of the Sarló-hát spectrum and beyond, the first copied from Fig. 4. Deviation graphs: 15 palynomorph taxa. Horizontal scale TBP ^{14}C yr; vertical scale right compositional transition acceleration *A;* vertical scale (left) deviations from random expectation; zero level on vertical scale (left) random expectation; positive deviation, over-performance relative to random expectation; negative deviation, under-performance. Selection criteria of taxa: 15 highest sum of squared deviations. Note: diploxylon refers to the pines with two fibrovascuar bundles, such as *Pinus sylvestris.*[19]

Having presented the graphs we have to re-emphasize that they are showing deviation. They are not distribution graphs in the usual sense since absences will also show up as deviation other than zero.

The objective of the remaining part of this chapter is to give an example of how we interpret the taxon deviation graphs (Fig. 12) alongside the *A,T* oscillation graphs:

i. We use 15 taxa of the available 122 in the palynomorph collection. The choice is guided by the total sum of squared deviations (TSSD) accounted for by the individual taxa. This is a practical way of choosing taxa if there is no *a priori* commitment to specific palynomorph types. The sum of

---

[19] Diploxylon or "hardwood" pines contrast with the haploxylon or "softwood" pines, such as the *Pinus strobus,* which have a single fibrovascular bundle. See http://en.wikipedia.org/wiki/Pinus_classification_for further details.

squared deviations SSD can be regarded as a measure of the information conveyed by the taxon in a physical sense or the taxon's characteristic (indicator) value in an ecological sense:

#	Taxon	SSD	% SSD/TSSD	PV	% PV
11	Corylus	141625.00	18.97	1529.70	24.87
38	Poaceae	133655.90	17.90	1468.58	23.88
14	Fagus	89853.64	12.03	781.26	12.70
5	Quercus	86601.99	11.60	653.36	10.62
13	Carpinus	52649.46	7.05	439.95	7.15
1	Pinus (diploxylon type)	47941.74	6.42	273.36	4.44
78	Umbelliferae	40284.78	5.40	293.54	4.77
8	Ulmus	34313.67	4.60	106.67	1.73
44	Artemisia	32163.93	4.31	272.82	4.44
61	Anthemis like	12073.87	1.62	57.28	0.93
16	Alnus glutinosa	11429.17	1.53	72.93	1.19
60	Compositae	10582.29	1.42	79.27	1.29
15	Betula	10121.02	1.36	45.33	0.74
10	Fraxinus excelsior	7213.96	0.97	24.52	0.40
45	Chenopodiaceae	7036.66	0.94	50.93	0.83

The # column in the table identifies taxon position in the original data set. The 15 taxa accounts nominally for about 96% of the total sums of squares. An alternative to ranking by the sums of squared deviations is ranking by the partial variance PV specific to the individual taxa. We compute the PV values by way of a series of orthogonal functions (Orlóci 2010). Note: in the table the SSD/TSSD% and PV% sequences are functionally correlated $(r = 0.9806)$.

ii. After explaining what the numbers signify, our first observation is that the floristic regions of the plain, the piedmont, the Carpathian Mountains, the major azonal floodplains, and other wetlands represent rich sources for palynomorph taxa in the Sarló-hát spectrum. It is quite obvious from the spectrum that the sources' pattern, zonal and azonal, are static only over very limited periods of time. In the long-run, the sources undergo transitions in situ and become historic stages in the long-term process.

iii. The Sarló-hát compositional transition process is mapped by the trajectory in Fig. 4 and *A* graphs in Fig. 6. Note the alternation of low and high instability phases. The high stability phases coincide with the flat segments (**B, D, F, H**) and the 'hotspots' or low stability phases coincide with the segments of violent oscillations (**A, C, E, G**). The following table has the cells dotted in the segments where a taxon's performance exceeds random expectation:

#	Taxon	A	B	C	D	E	F	G	H2	H1
11	Pinus	•								•
15	Betula	•	•							•
10	Fraxinus					•	•	•		•
11	Corylus			•	•	•	•	•		
8	Ulmus				•	•	•	•		•
5	Quercus			•	•	•	•	•		
45	Chenodiaceae	•	•				•	•	•	
44	Artemisia	•	•				•			•
78	Umbeliferae	•	•			•				
13	Carpinus	•	•	•						
16	Alnus	•	•	•	•	•	•			
14	Fagus	•	•							
38	Poaceae	•								•
16	Anthemus	•	•	•	•					
60	Compositae	•								

Based on the position of the dots in the table and based on the graphs in Fig. 12 we can construct a proxy vegetation history for of the Holocene Epoch on the NE Plain:

iv. The colossal temperature rise[20] between the ending of the last glaciation around the time of the infixion point in *T* (16890 BP (in Fig 12) is all

---

[20] The records in the Vostok series indicate a 4.62 °C temperature rise in the inversion layer over a 1115 yr period after the last flat section of the Vostok graph yr period 12261 BP until 11146 BC where our data SH data set begins. This must have materialized as a

but history by the time point 11131 BP where the Sarló-hát spectrum begins. By that time the early Postglacial vegetation is replaced by a new vegetation formation in which Pinus, Betula, Artemisia, Poaceae, and Ulmus perform much better than expected[21]. This transition from one vegetation formation to another opens up an era of relative compositional stability (period H1) which lasts at least 2500 years.

v. A new era begins with period **H2** characterised by the better than expected performance of Corylus and Quercus. By the time period **D** is reached, the Quercus-Corylus era is ending and a new formation starts assembling, highlighted by the better than expected performance of the Carpinus and Fagus palynomorphs.

vi. Period **A**, the most recent in the series, is transitional climatically. Interesting to observe the palynomorph mix contain Pinus, Betula, Umbeliferaea, Chenopodium, Artemisia, and Poaceae performing better than expected in association with Carpinus, Fagus, Alnus and Anthemis

vii. Quercus presents us with a unique history of dominance through the middle period from **H1** to **C**. It attains peak performance in the 'h' periods up to about 1900 BP after which its rapid decline is halted only for short periods. Of all species, Quercus suffered probably the most under the axe, typical for a taxon of high utility. Its rapid decline is well within the turbulent historic era (**A**) that saw the arrival of Huns and Magyars on the plains, introduction of massive grazing, and extensive flood controls

---

much larger temperature rise at ground level on the NE Plain, considering its Northern latitude and position in the hearth of a vast continent. Taking a proxy rate of 1.5 for latitude (Orlóci 2008) and considering a conversion rate of 2 from Vostok to global average temperature, we estimate the actual rise on the NE Plain to have amounted during the same 1115 yr period to $2 \times 1.5 \times 0.41/11.15 = 1.23$ °C per 100 years within the same 1115 yr period. All Vostok temperature values are differences in Celsius degrees and as such they do not have to be converted to the Kelvin scale before performing arithmetic operations on them.

[21] This mix of taxa is characteristic for the lower cold temperate portion of the current Spruce dominated circumpolar Boreal forest belt.

along the course of the Tisza river since the 19th Century. The end result is the demise of the Mighty Oak Savannah.

viii. The Poaceae undertakes decline following the **H** period, but separates out during period **A** with performance far better than expected. Carpinus stands solitary with its unimodal graph peaking around 3500 BP at the end of an 'h' period and a considerable dip in the *T* graph. It is likely that Carpinus became very common in the piedmont at that definitely turbulent time.

ix. The same may apply to the better than expected performance of Fagus during the **A,B** periods. Fagus and Carpinus, and also Alnus, Betula, and Pinus may have to be counted as palynomorph materials from other vegetation formation or azonal environments, and Quercus counted as if its recovery around 1900 BP continued into the **A** period. After these modifications, the hypotheses of the Oak savannah as the potential zonal vegetation formation could be revived and further tested (Fekete et al. 2008, 2010).

Process shape complexity

We are interested to estimate the shape complexity of the process trajectories. Our reason is in this to gauge the total random effect. The preferred complexity parameter is Mandelbrot's fractal dimension. We define the fractal dimension of the trajectory's mapping in the same manner as Mandelbrot (1967) did it for the English coast line. Orlóci (2010) summarised the theory and describes the *modus operandi* in an ecological context. Results are presented for Fig. 3 in Table 4 and Fig. 13.

Table 4. Shape complexity parameters of the three trajectories in figure 3. Legend: *n* number of length measurements in the set; *L(r)* graph length in units of the calliper width *r*; *b* linear regression coefficient of ln *L(r)* on ln *r* (see regression equations in Fig. 13); *D* fractal dimension. Definition of fractal dimension: $D=1-b$. Sarló-hát

	r	*L(r)*	*n*	*B*	*D*
Báb-tava	1	1689			
	2	1632	2	-4.95E-02	1.049528
	4	1460	3	-0.1051	1.105101
	8	1192	4	-0.1669	1.166903
	16	992	5	-0.19888	1.198877
	32	512	6	-0.31592	1.315915
	64	320	7	-0.39652	1.396517
	r	*L(r)*	*n*	*B*	*D*
	1	2325			

2	2302	2	-1.43E-02	1.014343
4	2232	3	-2.94E-02	1.029447
8	2008	4	-6.79E-02	1.067897
16	1728	5	-0.10534	1.105338
32	1216	6	-0.17341	1.173411
64	576	7	-0.29464	1.294643

Taul Dientre Brazi

r	L(r)	n	B	D
1	3582			
2	3536	2	-1.86E-02	1.018647
4	3460	3	-0.025	1.024997
8	3288	4	-4.02E-02	1.040201
16	2656	5	-9.68E-02	1.096793
32	2176	6	-0.14022	1.140217
64	1472	7	-0.20112	1.20112

The analytically determined (true) trajectory lengths in full dimension are 1689 (Sarló-hát), 2325 (Báb-tava) and 3582 (Taul Dientre Brazi) in units of the 226, 134 and 116 dimensional taxon based reference spaces.

The trajectory fractal dimensions rounded to 2 decimals in the same order: $D = 1 - b = 1 - (-0.396517) = 1.40$, 1.30 and 1.20 (for calliper width range from 1 to 64 in each case).

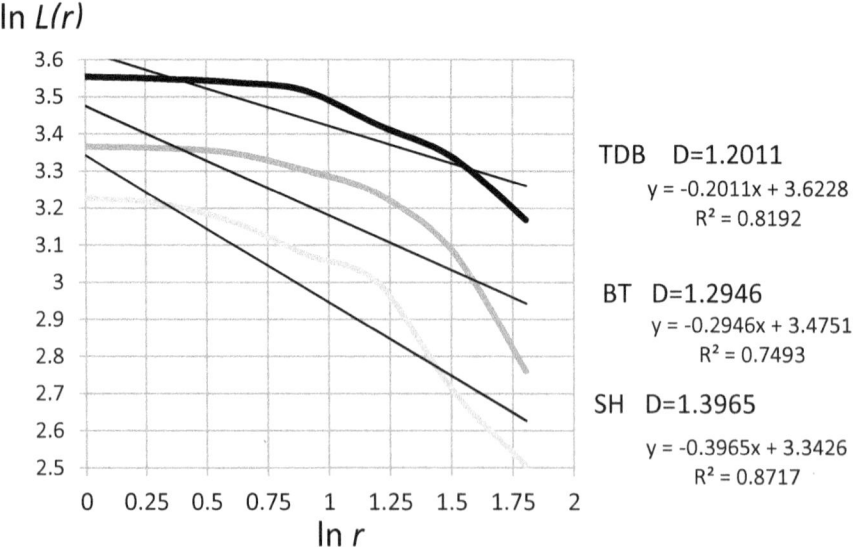

ln *L(r)*

TDB D=1.2011
y = -0.2011x + 3.6228
R² = 0.8192

BT D=1.2946
y = -0.2946x + 3.4751
R² = 0.7493

SH D=1.3965
y = -0.3965x + 3.3426
R² = 0.8717

ln *r*

74

Fig. 13. Graphical basis of the fractal dimension calculations for the Taul Dientre Brazi (top), Báb-tava (middle), and Sarló-hát (bottom) trajectories (Fig. 3). The graphs are based on data in Table 4. Horizontal axis: calliper width in nats. Vertical axis left: line length in nats. In regression equation y=bx+c: *b* regression coefficient, *x* horizontal axis ln *r, c* intercept on vertical axis ln *L(r)*, R^2 coefficient of determination.

What can we say about a trajectories shape complexity based on the *D* parameters? Considering that *D* is measured on a scale from 1 (least complex) to 2 (most complex), and further considering that *D* is reaching its theoretical upper bound (2) under the rule of chance, such as in the Brownian particle trajectory, none of the three observed *D* values should be considered extreme. This is interpretable as a sign that the trajectories' directed component is strong. In other words, the trajectory maps are not haphazardly laid lines within the phase space, but the traces of a well-defined deterministic process.

Comparison of trajectories

The trajectory graphs in Fig. 3 capture the processes in ways that make them directly comparable based on their homeomorphy. But to implement the comparisons in statistical terms, we need a special technique. The technique should answer the question: *Are the overlapping segments of the three trajectories homeomorphic to a degree to be called, by using the von Post terminology, parallel.*

The principal element of the technique is the topological coefficient $TC_\delta = \dfrac{M_\delta}{n}$. We already explained TC_δ earlier in the text and gave relevant references. We repeat, M_δ is a count of the number of times the next compositional state (the targeted trajectory point) moves away or toward the relevant reference state at tolerance radius δ in tandem of the compared trajectories. Symbol n is the total number of point pairs compared. We already mentioned too that:

i. The time scale is comparable in all cases.

ii. The level of homeomorphy is measured without direct reference to the shared identity of individual taxa.

iii. TC_δ is the scalar of multiscale process homeomorphy and not directly the compositional similarity of the palynomorph spectra in taxon terms. Two curtailed versions of the Sarló-hát trajectory are presented in Fig. 14. The curtailed versions have matching the time steps and time period in the Báb-tava and Taul Dientre Brazi trajectories.

Sarló-hát 1131 to 7781 c yr BP (134 steps)

Sarló-hát 9931 to 11287 c yr BP (29 steps)

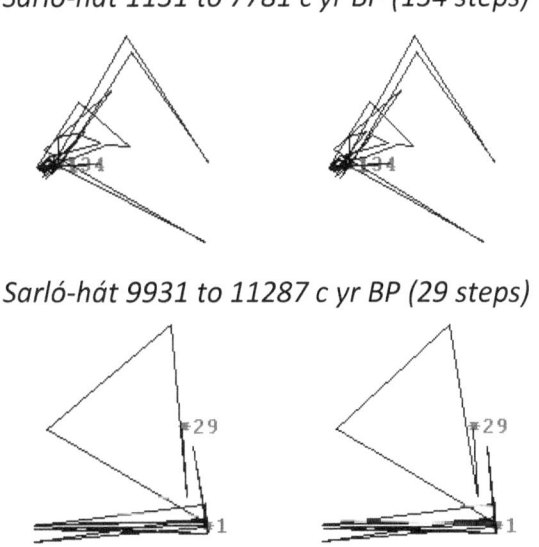

Fig. 14.The Sarló-hát trajectory shortened and redrawn in stereo dimensions to match time step and period length in the Báb-tava and the Taul Dientre Brazi trajectories of Fig. 3.

We present results in Fig. 15 and in the attached tables. Each comparison is associated with two diagrams. The graphs in the first diagram display the topological coefficient *(TC)*, expectations *(e)* and confidence limits *(UL, LL)*. The second graph displays the *TC - e* deviations. Except the *TC* values, the others are determined in a Monte Carlo experiment.

A. Sarló-hát to Báb-tava comparison

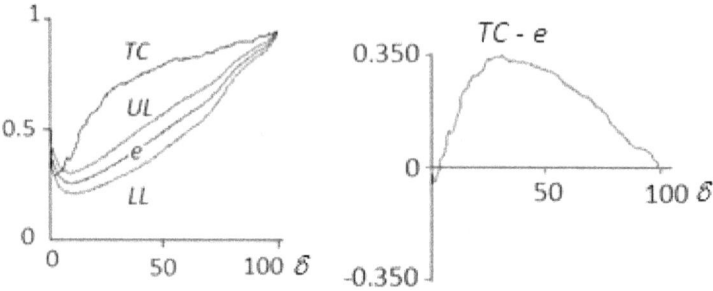

Statistics at given tolerance radius δ for significant *TC - e* at *L=0*

	δ	*e*	*Variance of e*	*LL*	*UL*
Maximum TC - e:	31	0.3563	0.0014	0.2836	0.429
Minimum TC - e:	6	0.2705	0.0005	0.2249	0.3161
		TC	*TC - e*	*(TC - e)/TC*	*P*
		0.7067	0.3503	0.9832	0.001
		0.3208	0.0502	0.1856	0.017

B. Sarló-hát to Taul Dientre Brazi comparison

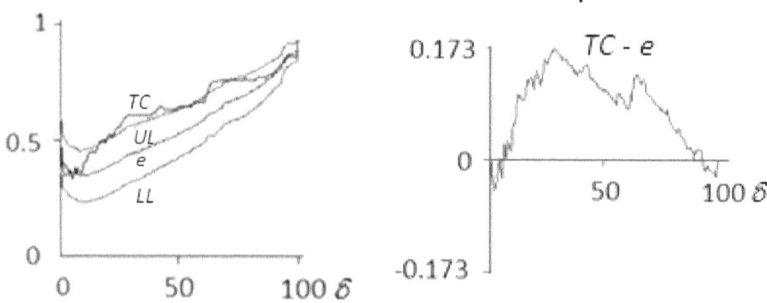

Statistics at the given δ for significant *TC - e* at *L=0*

	δ	*e*	*Variance of e*	*LL*	*UL*
Maximum TC - e:	28	0.4342	0.0042	0.3096	0.5587
Minima TC - e:	12	0.3501	0.0036	0.2348	0.4653
		TC	*TC - e*	*(TC-e)/e*	*P*
		0.6071	0.1729	0.3982	0.006
		0.4523	0.1022	0.2919	0.04

Fig. 15. Comparison of the Sarló-hát, Báb-tava, and Taul Dientre Brazi trajectories. Tabulated values correspond to the tolerance radius δ where the deviation of *TC* from random expectation *e* is maximal. Legend to symbols: *TC* – topological coefficient with values ranging from 0 to 1 (vertical axes); δ – tolerance radius ranging from 1 to 100 % (horizontal axis) based on the maximum distance of points on the trajectory; *e* – random expectation for *TC*; *UL, LL* – upper and lower 95% confidence limits about expectation *e*; *P* (in the tables) – the probability of a more extreme *TC* under random expectation. The *e, UL, LL,* and *P* values were determined in Monte Carlo experiments. Any observed value outside *UL* and *LL* is considered statistically significant.

A *TC* curve is interpreted in comparative terms. The reference for this is the *e* curve and also the *UL* and *LL* probabilistic limits. We mention again that the *e, UL* and *LL* curves are generated in a Monte Carlo experiment under the regularity condition that the compared palynological trajectories are the product of a completely random process. This is a hypothetical condition. When we find that *TC* - *e* exceeds a critical probability point marked by the confidence limits then we declare the deviation statistically significant at that probability or better, and at that value of δ.

For the expectation, we have a level of homeomorphy exactly 0.5 at tolerance radius δ=0. The homeomorphy of two trajectories can be more or it can be less than the expected. The ratio $HC = \dfrac{TC - e}{e}$ expresses relative homeomorphy from 0 to 1..

The topological coefficient *TC* merits further comments. When the trajectories are identical, the *TC* graph is a horizontal line across the top of the diagram at *TC* = 1. When the trajectories are independent, the *TC* graph is settled on or closely twined around the *e* graph. Any portion of the *TC* graph outside the *UL* and *LL* confidence limits is considered statistically significant at or better than the given probability. Conversely, when an observed *TC* graph is fully included within *UL* and *LL*, the compared processes are considered to lack any level of homeomorphy in excess of random expectation.

When a significant maximum *TC - e* occurs near zero on the δ axis, the trajectories are crisply homeomorphic with low random components in the probabilistic error sphere that surrounds the trajectory points. Conversely, the random component increases and homeomorphy is increasingly blurred when *TC - e* has maximum further out on the δ axis. As seen from the graphs, the homeomorphy of the Sarló-hát and Báb-tava trajectories is moderately crisp, reasonably strong, and statistically significant (within the matching time period). These conditions do not hold quite that well for the Sarló-hát and Taul Dientre Brazi comparison for which our calculus revealed no significant homeomorphy with a few isolated and numerically very weak exceptions.

Overview

The Essay clarifies the theoretical points and presents a case study which traces step by step the method of statistical multiscaling in a paleoecological environment. The objects are long-term processes manifested in the compositional transitions of palynological spectra.

In the multiscale environment, scale becomes a variable. Time step width, sample size, block size, confidence limits, tolerance radii, and time lag are examples for this in the essay. As a consequence, the results of the analysis are no longer single scalar quantities; they are scalar vectors with elements specific to the type of scale the analysis used.

This raises the problem of choice of a single, characteristic scalar out of the many to characterize the process in terms that are simple to view and easy to understand. The single scalars we chose are parameter maxima.

Another unusual but beneficial aspect of the analysis is the choice between generically different parameter functions from different algebras.

With the help of these, we capture trajectory characteristics such as Newtonian motion dynamics, Mandelbrot type fractal dimension, and in trajectory comparisons, the Eulerian homeomorphy. A constant feature in all of these is the statistical use of the parameters by evaluating the results with reference to probabilities, probabilistic limits, and expectations determined in Monte Carlo experiments.

Multiscaling allowed us to recognize complex, hierarchically embedded structures in the palynological spectra of sites in the NE portion of the Hungarian Great Plane and neighbouring Carpathian Mountains. Our results show clearly the scale dependence of the vegetation process and its fractal nature. Typical of this is the embedding of high and low stability process phases linked statistically to delayed temperature effects. We found indications of moderate process shape complexity, which indicates an overall strong process determinism, and different levels of homeomorphy, high for the SH and BT trajectory pair and low for the SH and TDB trajectory pair.

What did we find out that is new regarding the Holocene history of NE Hungary? Our results refine temporal limits in climate change, locate historic hotspots in compositional transitions, uncover definite, but delayed climatic effects, establish the presence of process homeomorphy, and all these in a statistical context.

Two interesting points mentioned in the text, but not discussed at any length, have to be addressed. These concern the justification for the use of the Vostok T oscillations as proxy for historic temperatures oscillations at the local sites and the use of linear scalars $r(A,T)$ and $\rho(A,T)$ to measure the correlation of the A and T series:

i. Regarding the Vostok T series, the source may appear geographically too remote to be relevant in NE Hungary. But we believe the Vostok T is oscillating in synchrony with the global warming and cooling cycles. We also believe that the regional climate cooling and warming cycles are manifestations of the global cycles. These not negate the fact that the magnitude of the temperature amplitudes realizing in different sites is region specific (Orlóci 1994, 2008). Under continental conditions in the Northern Hemisphere the amplitude increases with latitude.[22] This explains to us that while the global mean temperature rise may hardly be detectable by the available meteorological networks, climate warming already has definite manifestations in the Arctic and also in the Antarctic. Regarding NE Hungary, the use of the Vostok temperature series as proxy did bring forth meaningful results in multiscale correlation analysis.

ii. We seeded the text with bitts of information that point to the answer to why do we use a linear model such as $r(A,T)$ and $\rho(A,T)$, and why not some nonlinear contraption to measure the correlation of the A and T series? The answer is very simple: nonlinear models give results depending on the type of function or the applied polynomial's order. This is not a usable feature in our case, because it does not offer a unique choice.. We can improve the polynomial's fit and show higher and higher correlation with no end, simply by increasing the order of the polynomial. If not sooner, at order 20 for example, the Chebyshev polynomial easily gives a coefficient of determination (R^2) close to unity for any pair of data vectors. The linear model cannot be so manipulated and for that reason it provides a unique basis in comparative correlation studies.

Finally, I have to mention the code which I used to compute the various procedures. I have written these to support me, in my research. They are my tools. For application they require fluency in the techniques I use and

---

[22] Orlóci's function for estimation of thermal flax rate TFR at latitude L° in eastern North America is TFR=$0.0866L^\circ$- 2.3681 . This gives us

Latitude L°	80	70	60	50	40	40	30
TFR	4.559	3.693	2.827	1.961	1.095	1.095	0.229

Multiply this by the increase in the global mean temperature to obtain the actual thermal flux at any latitude.

skills to handle the many options. Once mastered, the numerous run-time options they offer are very effective in the handling of exploratory analyses of the data. Their initial use should therefore be in the course of expert tutorials. Application versions of the code are provided free of charge with Orlóci's "Statistical Ecology" and its external appendix "Flexible computing in Statistical Ecology" (2011).[23]

---

[23] Please visit https://sites.google.com/site/statisticalecology/

Bibliographic references

Borhidi, A., 1961: Klimadiagramme und klimazonale Karte Ungarns. - Annal. Univ. Sci. Budapest, Sect.Biol. 4: 21-50.

Braun-Blanquet, J. 1927.Pflanzensoziologie. -- 1932. Plant Sociology. (Translated by G.D. Fuller and H.S. Conard.) McGraw-Hill, New York

Clements F.E. 1916. Plant Succession: an Analysis of the Development of Vegetation. Publ. No. 242, Carnegie Institution, Washington.

Clements, F.E. 1928. Plant Succession and Indicators. H.W. Wilson Co., New York.

Conard, H.S. 1951. The Background of Plant Ecology. The Iowa State University Press, Ames (1977, Arno Press, New York).

Fekete G. , Molnár Zs., Kun A., Somodi I. and F. Horváth. 2008. Szárazgyepfajok a Duna-Tisza közén: elterjedési típusok és flóragrádiens. - In: Kröel-Dulay, Gy., Kalapos,T. Mojzes , A. (szerk.). Talaj-vegetáció-klíma kölcsönhatások. Köszöntjük a 70 éves Láng Editet. MTA ÖBKI, Vácrátót, pp.: 11-21.

Fekete G., Somodi I. and Zs. Molnár. (2010). Is chorological symmetry observable within the forest steppe biome in Hungary? - A demonstrative analysis of floristic data. Community Ecology 11:140-147

Feoli, E. and L. Orlóci. 1985. Species dispersion profiles of anthropogenic grasslands in the Italian Pre-Alps. Vegetatio 60: 113-118.

Gleason H.A. 1926. The individualistic concept of the plant association. Bull. Torrey Bot. Club 53: 7-26.

Gleason, H.A. 1917. The structure and development of the plant association. Bull. Torrey Bot. Club 44: 463-481.

Greig-Smith, P. 1952. The use of random and contiguous quadrats in the study of the structure of plant communities. Annals of Botany 16: 293-316.

Greig-Smith, P. 1983. Quantitative Plant Ecology. 3rd ed. Blackwell Scientific, London.

Hammersley J.M and D.C. Handscomb. 1964. Monte Carlo Methods. Methuen, London.

Kerner von Merilaun, A. 1863. Das Pflanzenleben der Donauländer. Innsbruck.

Kershaw A.P. 1994. Pleistocene vegetation of the humid tropics of northeastern Queensland, Australia. Palaeogeography, Palaeoclimatology, and Palaeoecology 109: 399-412.

Magyari, E., Jakab, G., Rudner, E. and P. Sümegi. 2001. Palynological and plant macrofossil data on Late Pleistocene short-term climatic oscillations in NE Hungary. Acta Palaeobotanica Supplement 2: 491–502.

Magyari, E., Sümegi, P., Braun, M., Jakab, G. and M. Molnár.. 2001. Retarded wetland succession: anthropogenic and climatic signals in a Holocene peat bog profile from the NE Carpathian Basin. Journal of Ecology 89: 1019–1032.

Magyari, E. 2002. Climatic versus human modification of the Late Quaternary vegetation in Eastern Hungary. PhD Thesis. Department of Mineralogy and Geology, University of Debrecen, Hungary.

Magyari, E. 2002b. The Holocene expansion of beech (Fagus sylvatica L.) and hornbeam (Carpinus betulus L.) in the eastern Carpathian basin. Folia Historico-Naturalia. Musei Matraensis 26: 15–35.

Magyari, E.K., Jakab, G., Sümegi, P. and Gy. Szöőr. 2008. Holocene vegetation dynamics in the Bereg Plain, NE Hungary – the Báb-tava pollen and plant macrofossil record. ACTA GGM DEBRECINA, Geology, Geomorphology, Physical Geography Series, Debrecen V. 3, 33–50.

Magyari, E.K., Chapman, J.C., Passmore, D.G., Allen, J.R.M., Huntley, J.P. and B. Huntley. 2008b. Holocene persistence of wooded steppe in the northern Great Hungarian Plain. – Journal of Biogeography.

Mandelbrot, B.B. 1967. How long is the coast line of Brittain? Statistical self similarity and fractal dimension. Science 156: 636-638.

McIntosh, R.P. 1985. The Background of Ecology: Concept and Theory. Cambridge University Press, New York.

Orlóci, L. 1971. An information theory model for pattern analysis. Journal of Ecology 59: 343-349.

Orlóci, L. and V. de Patta Pillar. 1989. On sample size optimality in ecosystem survey. Biométrie-Praximétrie. 29:173-184

Orlóci, L. and M. Orlóci. 1990. Edge detection in vegetation: Jornada revisited. Journal of Vegetation Science 1:311-324.

Orlóci, L. 1993. The complexities and scenarios of ecosystem analysis. In: G.P. Patil and C. R. Rao, Multivariate Environmental Statistics, pp.421-430, North Holland/Elsevier, New York.

Orlóci, L. 1994. Global warming: the process and its anticipated phytoclimatic effects in temperate and cold zones. Coenoses 9: 69-74.

Orlóci, L. 2001. Prospects and expectations: reflections on a science in change. Community Ecology 2: 187-196.

Orlóci, L. 2001b. Pattern dynamics: an essay concerning principles, techniques, and applications. Community Ecology 2:1-15.

Orlóci, L., Pillar, V.D., Anand, M. and H. Behling. 2002. Some interesting characteristics of the vegetation process. Community Ecology 3:125-146.

Orlóci, L., Pillar, V.D. and M. Anand. 2006. Multiscale analysis of palynological records: new possibilities. Community Ecology 7: 53-68.

Orlóci, L. 2008. Vegetation displacement issues and transition statistics in climate warming cycle. Community Ecology 9: 83-98.

Orlóci, L. 2009. Multi-scale trajectory analysis: powerful conceptual tool for understanding ecological change. Frontiers of Biology in China 4: 158-179

Orlóci, L. and K.S. He. 2009. On governance in the long-term vegetation process: How to discover the rules? Frontiers of Biology in China 4: 557–568.

Orlóci, L. 2010. Statistical Ecology. The quantitative exploration of nature to reveal the unexpected. Scada Publishing, London. 398 p. -- Online Edition: https://createspace.com/3476529

Orlóci. L. 2010b. Flexible Computing in Statistical Ecology. Scada Publishing, London. 138 p. -- Online Edition:

https://createspace.com/3574792

Orlóci, L. 2011. Self-organization and mediated transience in plant communities. Scada Publishing, London, Canada. -- Online edition: https://createspace.com/3585127

Petite, J.R., Jouzel, D., Raynaud, D., Barkov, N.I., Barnola, J.M., Basile, I., Bender, M., Chappellaz, J., Davis, J., Delaygue, G., Delmotte, M., Kotlyakov, V.M., Legrand, M., Lipenkov, V., Lorius, C., Pepin, L., Ritz, C., Saltzmann E. and M. Stievenard. 2001. Climate and atmospheric history of the past 420,000 years from the Vostok Ice Core, Antarctica. Nature 300: 429-436.

Pillar, V. De Patta and L. Orlóci. 1993. Taxonomy and perception in vegetation analysis. Coenoses 8:53-66.

Poore, M.E.D. 1962. The method of successive approximation in descriptive ecology. In: Advances in Ecological Research. Vol. 1, Academic Press, New York. pp. 35-68.

Post, von L. 1946. The prospects for pollen analysis in the study of the Earth climatic history. New Phytologist 45:193-217.

Schweingruber, F.H. 1996. Tree Rings and Environment Dendroecology. Paul Haupt, Stuttgart.

Szász, G. and L. Tőkei. 1997. Meteorológia mezőgazdáknak, kertésze-knek, erdészeknek. Mezőgazda Kiadó, Budapest.

Trewartha G.T. 2001. Global Mechanism of UNCCD, Via del Serafico 107, 00142 Rome, Italy. Web address: www.gm-unccd.org /English/ Field/aridity.htm

Wildi, O. and L. Orlóci. 1987. Flexible gradient analysis: a note on ideas and an application. Coenoses 2: 15-19.

Zólyomi, B. 1936.The history of ten thousand years from pollen grains. (In Magyar). Természet Tudományi Közlöny 1961-62, Hungarian Academy of Sciences, Budapest.

Bibliographic supplements

Quantum analysis of primary succession: The energy structure of a vegetation nosere in Hawaii Volcanoes National Park

ored by Laszlo Orlóci FRSC

List Price: **$30.00**

6" x 9" (15.24 x 22.86 cm)
Black & White on White paper
54 pages

ISBN-13: 978-1492788997 (CreateSpace-Assigned)
ISBN-10: 1492788996
BISAC: Science / Life Sciences / Ecology

The book revisits the classical idea that the potential energy structure of primary succession is a seamless fusion of foot-prints specific to basic processes which operate on all scales – phylogeny, environmental mediation, and chance. The idea is tested in quantum analysis of a vegetation chronosere in Hawai'i Volcanoes National Park. How is the test constructed? What are the outcomes? What do the results tell about primary succession not already known from other sources? Stated in the briefest of terms the test re-quires temporal species performance data.

ORDER FROM CREATESPACE ESTORE:
https://www.createspace.com/4452597

Quantum analysis of primary succession

The energy structure of a vegetation chronosere in Hawai'i Volcanoes National Park

László Orlóci FRSC

Quantum ecology: Energy structure and its analysis

Authored by László Orlóci FRSC

List Price: $30.00

6" x 9" (15.24 x 22.86 cm)
Black & White on White paper
72 pages

ISBN-13: 978-1492183297
ISBN-10: 1492183296
BISAC: Science / Life Sciences / Ecology

Ecology joins forces with quantum theory on the pages of "Quantum Ecology" to create a holistic approach in energy studies.

The infusion of quantum theoretical principles allows the study focus of ecological energetics to shift from the conventional calorific (trophic) flow in ecosystems to the potential energy structure of the vegetation. The books contents cover the theory and techniques in a unique account centred on the energy equation. The equation's component terms define energy footprints specific to ecology's basic processes, such as historic phylogeny, current environmental mediation of transience, and chance. What gives practical value to quantum analysis is its ability to be parameterised by the usual type of survey or experimental data.

The book is offered for classroom use in advanced courses and technical support in research projects.

ORDER FROM CREATESPACE ESTORE:
https://www.createspace.com/4406077

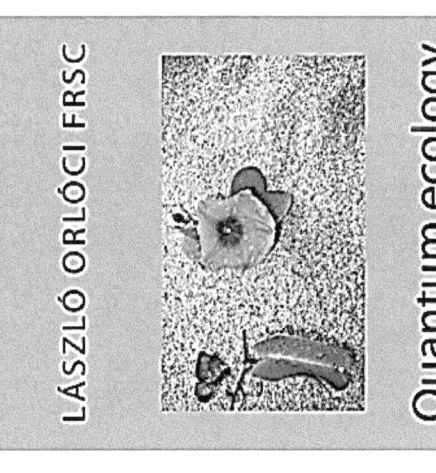

LÁSZLÓ ORLÓCI FRSC

Quantum ecology

Energy structure and its analysis

Statistical ecology

Statistical ecology

The quantitative exploration of Nature to reveal the unexpected

László Orlóci FRSC

The quantitative exploration of Nature to reveal the unexpected

Authored by Laszlo Orlóci Ph.D.

The book's topics traverse many problem areas in univariate and multivariate data analysis, focussed on current ecological practice. The manner of presentation emphasizes reasoned methodological choices and encourages innovations consistent with the objectives, but mindful of the need to see clearly the regularity conditions which set limits for valid application of statistics in Ecology. The main text is accompanied by external appendices including a technical manual, 47 specialized application programs, and many data files taken from the exercises in the main text. For information please contact: lorloci@uwo.ca

List Price: $49.90

Add to Cart

About the author:

Orlóci is an INTECOL Distinguished Statistical Ecologist. He is external (academician) Member of the Hungarian Academy of Sciences, and regular (academician) Fellow of the Academy of Sciences of the Royal Society of Canada. He published over 100 papers in scientific journals, numerous monographs and books. His current essays on trajectory analysis, the rules of process governance, and the phylogenetic signal in vegetation transitions have considerable significance for evolutionary ecology and global change science. His present work on energy structures in metacommunities is seminal, pointing to a new direction.

Publication Date:	Aug 10 2010
ISBN/EAN13:	1453760520 / 9781453760529
Page Count:	372
Binding Type:	US Trade Paper
Trim Size:	6" x 9"
Language:	English
Color:	Black and White
Related Categories:	Science / Life Sciences / Ecology

Statistical multiscaling in dynamic ecology

f Like 0

Probing the long-term vegetation process for patterns of parameter oscillation

Authored by László Orlóci Ph.D.

The Book's conceptualisation of multiscaling theory presents the Next Step in the study of the long-term vegetation process. The context is statistical and the process generating events have proxy in the compositional transitions of the palynological spectra. Familiarity with multiscaling is not a pre-requisite. The reader shall learn from the examples how multiscaling techniques helped to identify the self-similar (fractal) nature of the process, isolate low and high instability phases, locate hotspots of compositional transitions, and link these to delayed climatic effects. He or she shall also learn how to gauge process homeomorphy among sites, isolate the random and directed effects found braided into the process, and do much more within a broad yet formal probabilistic framework. The Book's contents are taken in part from a graduate course offered in the Ecology program at UFRGS in Porto Alegre, Brazil. The examples use palynological spectra from sites on the Hungarian Great Plain and in the adjacent Carpathian Mountains. Application programs are available from the author.

List Price: $30.00

[Add to Cart]

Publication Date:	Mar 15 2012
ISBN / EAN13:	1475071388 / 9781475071382
Page Count:	96
Binding Type:	US Trade Paper
Trim Size:	6" x 9"
Language:	English
Color:	Black and White
Related Categories:	Science / Life Sciences / Ecology

93

About the author:

László Orlóci was born in Hungary in 1932. He holds degrees in forest engineering (DFE Sopron), forestry science and biology (BSF, MSc, PhD University of British Columbia), and DSc h.c. in biology (University of Trieste). Orlóci held appointments as NATO Science Fellow (University College of North Wales), professor (University of Western Ontario), and visiting professor at universities in the Americas, the Pacific, Asia, and Europe. He is an INTECOL Distinguished Statistical Ecologists, external (academician) member of the Hungarian Academy of Sciences, and regular Fellow of the Academy of Sciences of the Royal Society of Canada.

Self-organization and mediated transience in plant communities

Like 0

What are the rules?

Authored by Dr. László Orlóci FRSC

A novel, signal theoretical solution is sketched out for the ecological problem of how to identify and quantitatively express the assembly rules of plant communities. A case study for testing the solution leads to the astonishing conclusion that the phylogenetic signal outperforms the current environmental signal in intensity close to 7 to 1. This indicates high stability and low inclination to environment mediated transience in the community.

List Price: $25.00

Add to Cart

Publication Date:	Nov 11 2011
ISBN/EAN13:	1461028221 / 9781461028222
Page Count:	70
Binding Type:	US Trade Paper
Trim Size:	6" x 9"
Language:	English
Color:	Black and White
Related Categories:	Science / Life Sciences / Ecology

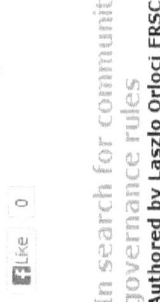

List Price: $30.00

Add to Cart

On the energy structure of natural vegetation

f Like 0

In search for community governance rules

Authored by Laszlo Orloci FRSC

Briefly about the book ...

Vegetation Science meets quantum theory in the energy-based entropy model of this book. The model is based on Max Planck's postulate that potential energy and entropy are interchangeable parameters in resonator complexes. What is a typical outcome of the model in vegetation studies? The model hands users a set of entropy estimates and probabilities based on which the strength and uniqueness of a metacommunity's energy structure can be characterised in comparative terms.

About the author:

Orlóci is an INTECOL Distinguished Statistical Ecologist. He is external (academician) Member of the Hungarian Academy of Sciences, and regular (academician) Fellow of the Academy of Sciences of the Royal Society of Canada. Orlóci published over 100 papers in scientific journals, numerous monographs, books and book chapters. His current essays on trajectory analysis, the rules of process governance, and the phylogenetic signal in vegetation transitions have considerable significance for evolutionary ecology and global change science. His present work on energy structures in metacommunities is seminal, pointing to a new direction.

Publication Date: Jan 30 2013
ISBN/EAN13: 1482319373 / 9781482319378
Page Count: 46
Binding Type: US Trade Paper
Trim Size: 6" x 9"
Language: English
Color: Black and White
Related Categories: Science / Life Sciences / Ecology

Flexible computing in statistical ecology  Like 0

External appendix to accompany L. Orlóci's Statistical Ecology

Authored by Dr. László Orlóci

Problem flexible computing in statistical ecology

The Book describes more than 40 executable (.exe) computer programs and presents examples of application which correspond to the examples included in Statistical Ecology*. The programs are flexibly problem specific and conversational. They allow option-driven selective access to specific statistical tasks. Linkages are possible through standard output and input. The description includes in each case a brief introduction, a record of the start up dialogue, and detailed record input and output sets. The source code is in True Basic. The programs are compiled and linked on a 32 bit Windows XP system and tested up to Windows 7.

The executable program library, data files and a note to users are distributed free of charge to purchasers of the Technical Manual. Requests for download information should be directed to the URL address lorloci@uwo.ca. Prior to running the application programs, installation of a recent version of True Basic (see Internet for sources) on the user's system is strongly advised.

* Orlóci, L. 2010. Statistical Ecology. The quantitative exploration of nature to reveal the unexpected. Scada Publishing, Online Edition. Copies are available from the distributor
https://www.createspace.com/3476529

Publication Date:	Apr 05 2011
ISBN/EAN13:	1460972953 / 9781460972953
Page Count:	142
Binding Type:	US Trade Paper
Trim Size:	6" x 9"
Language:	English
Color:	Black and White
Related Categories:	Science / Life Sciences / Ecology

List Price: $30.00

Add to Cart

Statistical Ecology. A reasoned approach.

Index

A and T series, 48, 53
A graphs, 26
A oscillations, 39
absolute dominance, 56
acceleration, 26, 39, 51, 82
acute angle, 39
algorithm for rescaling, 30
atmospheric aridity, 58
averaging, 42
Báb-tava, 15, 38, 74, 77, 78, 79, 80, 87
Black Sea, 16
block size, 30, 81
Borhidi, A., 16
Brownian particle trajectory, 75
Clements, F.E., 12
climate change, 57, 58, 82
comparability, 32
compositional transition, 26, 60
compositional transition acceleration, 12
Conards, 12
confidence limits, 79
correlation, 40, 43
data smoothing, 30
deceleration, 26, 39
delayed synchrony, 13, 23
deuterium, 17
dimension, 73
e, 35, 78, 79
ecology, iv

Eigenanalysis, 26, 38
embedded graphs, 82
English coast, 73
error control, 43
Eulerian topology, 12
F^+, 43, 44
forcing factors, 26
fractal, 26, 73, 74
fractal dimension, 39, 75, 77
fractaldimension, 14
F^-, 43, 44
Gleason, H.A., 12
Greig-Smith, 86
Greig-Smith technique, 11
Greig-Smith, P., 10
Hawaii, 14
He, 101
Holocene, 13, 16, 57, 58, 82, 86, 87
homeomorphy, 12, 24, 26, 33, 79, 80
hotspots, 13, 23, 26, 42, 82
humid climate, 56
Hungary, 15, 16, 82, 86, 87
Huns, 71
inner angle, 39
Kailua, 14
Kerner, 12, 86
L, iv, 43, 44, 48, 53, 54, 55, 56, 57, 59, 60, 78, 86, 87, 88
lag, 36, 45, 53, 55, 56, 57

loss of information, 32
Magyar tribes, 17
Magyari, v, 13, 15, 16, 21
Magyari, E.K., v, 13, 16, 17, 86, 87
Magyars, 71
Mandelbrot, 73
Mandelbrot fractal, 12
McIntosh, 87
McIntosh, Robert P., 12
modus operandi, 8, 73
Monte Carlo, 19, 36, 49, 50, 78, 79, 82, 86
Monte Carlo experiment, 36, 79
moving averages, 30
moving window, 42
MTA, v, 13
multiscale, 25, 42
multiscale statistical analysis, 14
multiscaling, 8, 9, 10, 11, 12, 13, 14, 15, 19, 20, 28, 82
Newtonian, 82
Oak savannah, 72
Orlóci, iv, 10, 11, 14, 16, 17, 18, 19, 25, 27, 28, 29, 42, 44, 49, 52, 53, 57, 62, 69, 71, 83, 84, 86, 87, 88, 89, 101
Orlóci, L., iii, 17, 18, 25, 32, 42, 51, 62, 73, 87, 88
Orlóci,L., 28
over-performance, 62, 68
Oxbow landscape, 9
paleoecology, 13
palynological, 11, 14, 15, 16, 17, 19, 20, 21, 25, 32, 37, 51, 79, 81, 82, 87

palynological spectra, 20, 26, 32, 51
parallelism, 23
parameter vector, 29
parameters, 12, 26, 29, 58, 73, 75
partial variance, 11, 69
particle identification, 18
performance parameters, 11
Petite, J.R., 17
polar co-ordinates, 27
Pontic steppe, 16
Poore, 8, 88
Poorean successive approximation, 28
power of resolution, 9, 28, 29
Principle Components Analysis, 26
probability, 48, 79
process, 23, 25, 26, 33, 35, 37, 51, 57, 60, 87
process determinism, 14
process phases, 13
Quantitative Plant Ecology, 11, 86
Quercus, 71
r(A,T), 42, 43, 44
random effect, 23, 26, 42
random expectation, 35, 68, 79
recursive application, 29
regression, 42, 44, 54, 55, 56, 73, 75
Retezat Mountains, 15
Romania, 15
sampling error, 18
Sarló-hát, 15, 40, 56, 58, 62, 74, 77, 78, 80

scale effect., 9
scales, 29
scaling up, 32
shape complexity, 12, 73, 75
Sharló-hát, 38, 43, 79
shifting the record series, 36
slope, 55
spatial patterns, 26
statistical significance, 50
stereograms, 38, 39
stretched records, 30
Swinegruber, F.H., 17
Szász, G., 16
Taul Dientre Brazi, 15, 38, 74, 77, 78, 79, 80
Taul DientreBrazi, 74, 75
taxa, 13, 15, 16, 20, 21, 26, 32, 33, 38, 39, 61, 62, 68, 69, 71, 77
TC graphs, 26
telemetric data, 9
temperature oscillations, 13, 51
Thornthwait index, 56, 58
Time step, 81
time step ransformation, 32
time step width, 30

Tisza floodplain, 9
Tiszafüred, 9
Tőkei, L., 16
tolerance limit, 32, 34, 35, 79, 81
topological coefficient, 35, 76, 77, 79
Török, K., v
trajectory analysis, 25, 26, 29
trajectory lengths, 74
trajectory map, 26
trajectory point, 39, 76
Trewartha, 89
under-performance, 62, 68
vegetation, iv
vegetation transitions, 13
von Post, 76
von Post doctrine, 23
Vostok, 88
Vostok temperature, 17, 40, 51
Wildi, 28
Zólyomi, v, 16, 89
Φ^-, 43, 44, 48, 51, 54, 55, 56, 58, 59
$\rho(A,T)$, 48, 51, 54, 55, 56, 58, 59, 82

Biographical notes

 László Orlóci was born in Hungary in 1932. He holds degrees in forest engineering (DFE Sopron), forestry science and biology (BSF, MSc, PhD University of British Columbia), and DSc *h.c.* in biology (University of Trieste, Italy). Orlóci held appointments as NATO Science Fellow (University College of North Wales), professor (University of Western Ontario), and visiting professor at universities in the Americas, the Pacific, Asia, and Europe.

Orlóci is an INTECOL Distinguished Statistical Ecologists, external (academician) member of the Hungarian Academy of Sciences, and regular Fellow of the Academy of Science of the Royal Society of Canada. He published over 100 papers in scientific journals, numerous monographs, book chapters and books. His current essays on multiscale analysis have considerable utility in dynamic and global change ecology.

 Orlóci is married to Márta Mihály. Márta is a fellow Sopron forest engineering alumna (DFE, BSF), life-time research associate. Their daughter Martha Barbara is a Geography graduate of UWO. They have two granddaughters, Kathryn Orlóci-Goodison is taking Forestry in the Faculty of Natural Resources Management of Lakehead University in Thunder Bay, Ontario, and Ruth Orloci-Goodison is attending high school in Cambridge, Ontario.

Reader's notes

www.ingramcontent.com/pod-product-compliance
Lightning Source LLC
Chambersburg PA
CBHW051338170526
45166CB00002B/872

*9 7 8 1 4 7 5 0 7 1 3 8 2 *